Marine-Based Bioactive Compounds

The ocean offers boundless benefits to human health. Known for millennia as a source of food, it is continuously gaining recognition as a provider for a variety of materials, and as the largest habitat on our planet, the ocean's biodiversity stands far above anywhere else.

Functional ingredients derived from marine algae, invertebrates, and vertebrates can help fill the need for novel bioactives to treat chronic conditions such as cancer, microbial infections, and inflammatory processes. The latest addition to the *Nutraceuticals: Basic Research/Clinical Applications* series, **Marine-Based Bioactive Compounds: Applications in Nutraceuticals** provides an account of marine-derived nutraceuticals and their potential health benefits.

Key Features:

- Provides the latest information on trends in this fast-growing market
- Focuses on the six marine taxa that offer the most exciting potential
- Gives practical guidance to anyone wishing to enter this sector

With contributions from an international group of experts, this book presents a great many opportunities in marine nutraceuticals from the six oceanic taxa that offer the most potential to benefit human health. It is a great resource for established nutraceutical companies.

Nutraceuticals: Basic Research/ Clinical Applications

Series Editor: Yashwant V. Pathak, PhD

Marine Nutraceuticals: Prospects and Perspectives
Se-Kwon Kim

Nutrigenomics and Nutraceuticals: Clinical Relevance and Disease Prevention
edited by Yashwant V. Pathak and Ali M. Ardekani

Food By-Product Based Functional Food Powders
edited by Özlem Tokuşoğlu

Flavors for Nutraceuticals and Functional Foods
M. Selvamuthukumaran and Yashwant V. Pathak

Antioxidant Nutraceuticals: Preventive and Healthcare Applications
Chuanhai Cao, Sarvadaman Pathak, and Kiran Patil

Advances in Nutraceutical Applications in Cancer: Recent Research Trends and Clinical Applications
edited by Sheeba Varghese Gupta and Yashwant V. Pathak

Flavor Development for Functional Foods and Nutraceuticals
M. Selvamuthukumaran and Yashwant V. Pathak

Nutraceuticals for Prenatal, Maternal and Offspring's Nutritional Health
Priyanka Bhatt, Maryam Sadat Miraghajani, Sarvadaman Pathak, and Yashwant V. Pathak

Bioactive Peptides: Production, Bioavailability, Health Potential and Regulatory Issues
edited by John Oloche Onuh, M. Selvamuthukumaran, and Yashwant V. Pathak

Nutraceuticals for Aging and Anti-Aging: Basic Understanding and Clinical Evidence
edited by Jayant Lokhande and Yashwant V. Pathak

Marine-Based Bioactive Compounds: Applications in Nutraceuticals
edited by Stephen T. Grabacki, Yashwant V. Pathak, and Nilesh H. Joshi

For more information about this series, please visit: www.crcpress.com/ Nutraceuticals/book-series/CRCNUTBASRES

Marine-Based Bioactive Compounds

Applications in Nutraceuticals

Edited by Stephen T. Grabacki,
Yashwant V. Pathak, and Nilesh H. Joshi

CRC Press
Taylor & Francis Group
Boca Raton London New York

CRC Press is an imprint of the
Taylor & Francis Group, an **informa** business

First edition published 2023
by CRC Press
6000 Broken Sound Parkway NW, Suite 300, Boca Raton, FL 33487-2742

and by CRC Press
4 Park Square, Milton Park, Abingdon, Oxon, OX14 4RN

CRC Press is an imprint of Taylor & Francis Group, LLC

Library of Congress Cataloging-in-Publication Data
Names: Grabacki, Stephen T., editor. | Pathak, Yashwant Vishnupant, editor. |
    Joshi, Nilesh H., 1975– editor.
Title: Marine-based bioactive compounds : applications in nutraceuticals /
    edited by Stephen T. Grabacki, Yashwant V. Pathak and Nilesh H. Joshi.
Description: First edition. | Boca Raton, FL : CRC Press, 2023. | Series: Nutraceuticals.
    Basic research/clinical applications | Includes bibliographical references and index.
Identifiers: LCCN 2022029004 (print) | LCCN 2022029005 (ebook) | ISBN
    9780367614935 (hbk) | ISBN 9780367651749 (pbk) | ISBN 9781003128175 (ebk)
Subjects: LCSH: Bioactive compounds. | Marine pharmacology.
Classification: LCC QP517.B44 M36 2023 (print) | LCC QP517.B44 (ebook) |
    DDC 572/.69—dc23/eng/20220916
LC record available at https://lccn.loc.gov/2022029004
LC ebook record available at https://lccn.loc.gov/2022029005

ISBN: 978-0-367-61493-5 (hbk)
ISBN: 978-0-367-65174-9 (pbk)
ISBN: 978-1-003-12817-5 (ebk)

DOI: 10.1201/9781003128175

Typeset in Garamond
by Apex CoVantage, LLC

To my loving wife Marianne Kerr—"side by side".

And to the memory of my parents—"roots and wings".

Stephen T. (Steve) Grabacki

To the loving memories of my parents and Dr. Keshav Baliram Hedgewar, who gave proper direction to my life, to my beloved wife Seema who gave positive meaning and my son Sarvadaman who gave a golden lining to my life.

I would like to dedicate this book to the loving memories of Ma Chamanlaljee, Ma Lakshmanraojee Bhide and Ma Madhujee Limaye who mentored me selflessly and helped me to become a good and socially useful human being.

Yashwant V. Pathak

I would like to dedicate this book to my respected teacher, late Dr. A. Y. Desai, whose constant guidance and commitment have always blessed my life.

Sir, I miss you more than words can say.

Nilesh H. Joshi

Contents

Preface

The world ocean offers limitless benefits to human health. Known for millennia as a source of food, the ocean is becoming recognized (and valued!) as a provider of a fantastic array of materials to the medical, fitness, and health and beauty community.

The ocean is, by far, the largest habitat on our planet, in both area and in volume. As such, its biodiversity stands head-and-shoulders above anywhere else. New species are being discovered all the time, from tidelands, polar seas, and the mesopelagic and abyssal zones among many others. Further, there is a virtual tsunami of discoveries of the bountiful contributions of those myriad organisms to human life, health, and well-being.

A nutraceutical is a substance which benefits human health, but which is not considered to be a drug (pharmaceutical). To guide the reader, and to illustrate the wide range of materials, here are two useful definitions of "nutraceutical"—

- A chemical substance or group of substances that for legal purposes is defined as a nutrient but that is in fact marketed and used for the prevention or treatment of disease. (Farlex Partner Medical Dictionary)
- A food or naturally occurring food supplement thought to prevent disease or have other beneficial effects on human health. Also called *functional food*. (The American Heritage® Medical Dictionary)

The global nutraceutical market is huge, and it's getting bigger all the time. The inter-related trends that drive this amazing growth include—

- Rise in chronic and non-communicable disorders: heart disease, diabetes mellitus, etc.

- Increasing health care costs: an almost exponential trend
- Major demographic segments: purchasing power of baby boomers and millennials
- Desire for natural products and remedies: non-prescription remedies
- Parallel trends: weight management and wellness-focused diets
- Do-it-yourself health care: people take health into their own hands
- Convenience and availability: information and purchasing via the internet

This book presents a great many opportunities in marine nutraceuticals from the six oceanic taxa that offer the most potential to benefit human health. We hope that you find it to be both useful and exciting.

Editors

Stephen T. Grabacki is President and owner of GRAYSTAR Pacific Inc., a consulting and R&D company in Anchorage, Alaska. For more than 40 years, he has provided scientific and technical support services to clients in the field of responsible development of marine-related natural resources—fisheries, minerals, and energy. Grabacki holds a Master of Science degree in fisheries biology (minor in business management) from University of Alaska Fairbanks. He is a Certified Fisheries Professional (American Fisheries Society), a recipient of the Antarctic Service Medal of the United States (with Winter-Over insignia), and an Eagle Scout. For four years, Grabacki served as Chairman of the Board of the Alaska SeaLife Center in Seward, Alaska, and remains an active member as Board Mentor (informally termed "Obi-Wan"). At University of Alaska Anchorage, he has taught courses in oceanography, fisheries management, seafood logistics, and seafood marketing. He is a frequent communicator on topics regarding ocean-related business development. He contributed two chapters to the CRC Press book *Environmental Effects on Seafood Availability, Safety and Quality Issues*.

Yashwant V. Pathak is an Adjunct Professor at Faculty of Pharmacy, Airlangga University, Surabaya, Indonesia and he has over 13 years of versatile administrative experience in an institution of higher education as Dean (and over 30 years as faculty and as a researcher in higher education after his Ph.D.). Dr. Pathak presently holds the position of Associate Dean for Faculty Affairs and Tenured Professor of Pharmaceutical Sciences at Teneja College of Pharmacy, University of South Florida. He is an internationally recognized scholar, researcher, and educator in the areas of health care education, nanotechnology, drug delivery systems, and nutraceuticals. He has published extensively with over 50 edited volumes to his credit in the areas of nanotechnology, drug delivery systems, artificial neural networks, conflict management, and cultural studies. Elsevier, John Wiley and Sons, Springer, and Taylor & Francis, are among the many international publishers publishing his books. He has published over 300 research papers, reviews, and chapters in the books and presented in many national and international conferences. He is also actively involved in many nonprofit organizations, including Hindu Swayamsevak Sangh, USA; Sewa International USA; International Accreditation Council for Dharma Schools and Colleges; International Commission for Human Rights and Religious Freedom.

Nilesh H. Joshi is an Associate Professor in the College of Fisheries Sciences, Kamdhenu University, Veraval, Gujarat, India. Dr. Joshi holds an M.Sc. and Ph.D. in Biological Science from Saurashtra University, Gujarat, India. For the last 16 years, he has been a focused researcher in the field of seaweed biology, biochemicals and phycocolloids, and seaweed cultivation of economically important species. He has documented and described more than 170 seaweed species from the Gulf of Kutch and has also worked on the assessment of the primary productivity of ocean waters using remote sensing techniques. At the College of Fisheries Sciences, he teaches basic biology, coastal aquaculture and ecology, limnology, and seaweed mariculture. Under the coastal community development program, he provides technical know-how on seaweed cultivation and its uses on a commercial scale in India. He has also contributed as an editor to the CRC Press book *Seaweeds as Plant Fertilizer, Agricultural Biostimulants, and Animal Fodder.*

Contributors

Gandhi Rádis-Baptista
Institute for Marine Sciences
Federal University of Ceara
CE, Brazil

Rajesh V. Chudasama
College of Fisheries Science
Kamdhenu University, Veraval
Gujarat, India

Lisa Fajar Indriana
Indonesian Institute of Sciences
  (LIPI)
West Nusa Tenggara, Republic of
  Indonesia

Se-Kwon Kim
College of Science & Technology
Hanyang University
Gyeonggi-do, Republic of Korea

Ilza R. Mor
H & H.B. Kotak Institute of Science
Saurashtra University, Rajkot
Gujarat, India

Khushali M. Pandya
The Maharaja Sayajirao University
  of Baroda
Padra, Vadodara, India

Ratih Pangestuti
Indonesian Institute of Sciences
  (LIPI)
West Nusa Tenggara, Republic of
  Indonesia

Idham S. Pratama
Indonesian Institute of Sciences
  (LIPI)
West Nusa Tenggara, Republic of
  Indonesia

Yanuariska Putra
Indonesian Institute of Sciences
  (LIPI)
West Nusa Tenggara, Republic of
  Indonesia

Puji Rahmadi
Indonesian Institute of Sciences
  (LIPI)
West Nusa Tenggara, Republic of
  Indonesia

Maya Raman
Kerala University of Fisheries and
  Ocean Studies
Kochi, Kerala, India

1

Nutraceuticals from Fishes

Stephen T. Grabacki

Contents

1.1 Introduction

Finfishes offer the most exciting and potentially profitable sources of marine nutraceuticals. Thus, it is appropriate that the first chapter in this book highlights the best range of combination science-plus-business opportunities in this field.

DOI: 10.1201/9781003128175-1

Fishes are uniquely rich sources of marine nutraceuticals for three reasons—(1) there are many available species, and each species has its own suite of biochemical substances (pharmaceuticals, nutraceuticals, and cosmeceuticals); (2) unlike most of the other taxa discussed in this book, many fish species are well-known, available in abundant quantities, and sustainable in supply; and (3) the seafood processing industry generates huge volumes of waste—which we call "byproducts"—and thus has a strong incentive to obtain value from that waste stream.

Fish-based nutraceuticals span a wide range of materials (Figure 1.1).

This chapter is written from the perspective of the seafood producing industry (both wild capture and aquaculture) because industry participants—

- Routinely harvest fish species which have interesting nutraceutical properties
- Have a strong economic incentive to obtain maximum value from those harvests

And the focus of this chapter is on finfish only. Other marine taxa—crustaceans, mollusks, etc.—are discussed in other chapters of this book.

The annual volume of commercial finfish harvests in the American Exclusive Economic Zone (EEZ) sustainably fluctuates around eight million pounds (Figure 1.2), but the value of those harvests has been increasing (Figure 1.3).

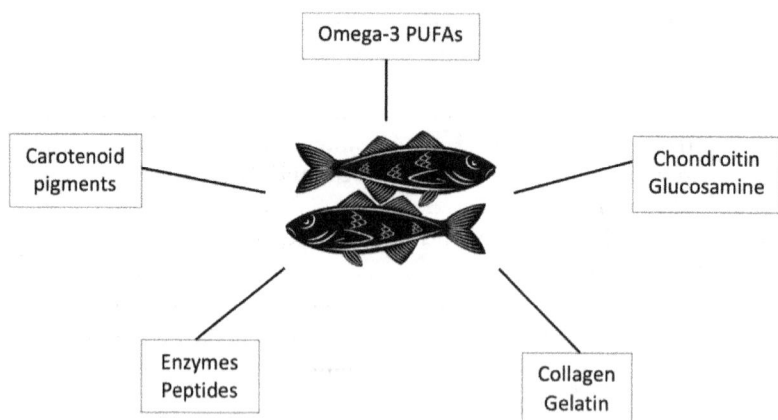

Figure 1.1 *Schematic outline of the major fish-based nutraceuticals.*

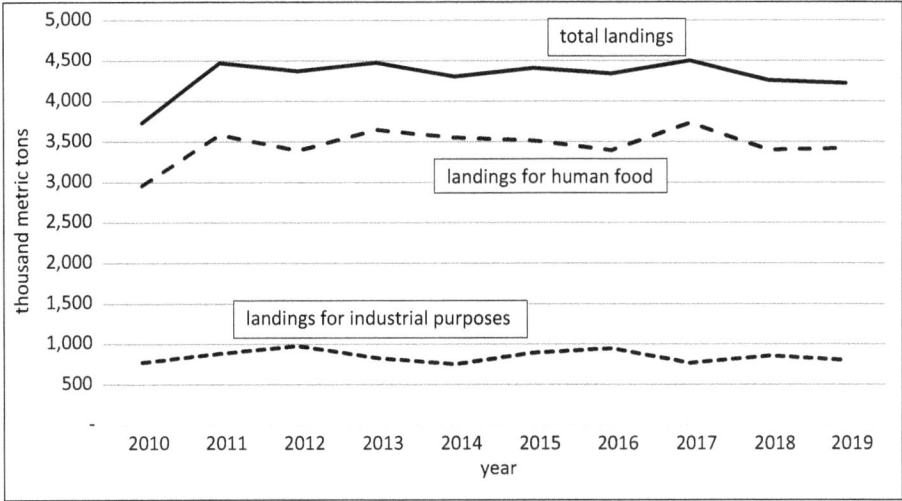

Figure 1.2 *Volume of US domestic finfish and shellfish landings, 2010–2019.*

Source: National Marine Fisheries Service (2021)

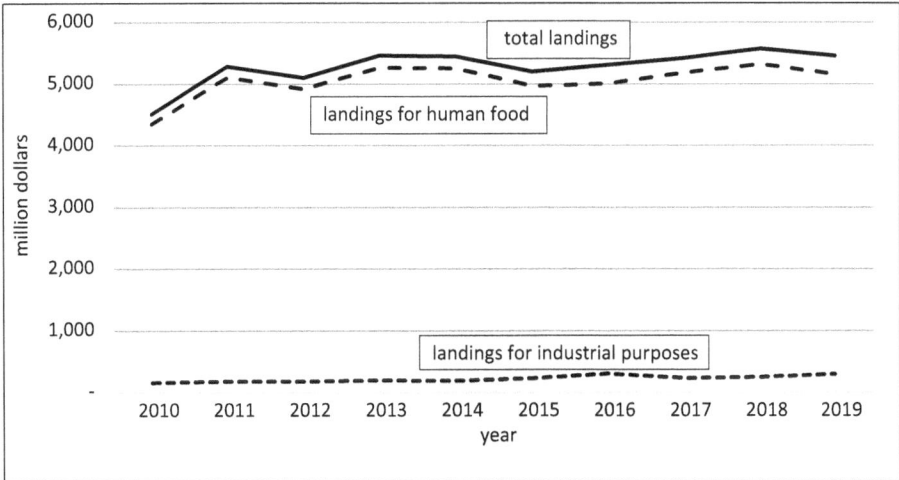

Figure 1.3 *Value of US domestic finfish and shellfish landings, 2010–2019.*

Source: National Marine Fisheries Service (2021)

Well over half of those harvests come from the US EEZ off Alaska—in 2018, Alaska produced 5.4 billion pounds, for about 57.4% of the total harvest in the entire American EEZ.

World harvests are, of course, far larger. In 2018, world aquaculture production and capture fisheries produced over 122.5 million metric tons of finfishes, most of which came from the Pacific and Atlantic areas (FAO 2020; Figure 1.4).

Most of those harvests are processed into frozen and "fresh" (i.e. chilled) products (Figure 1.5).

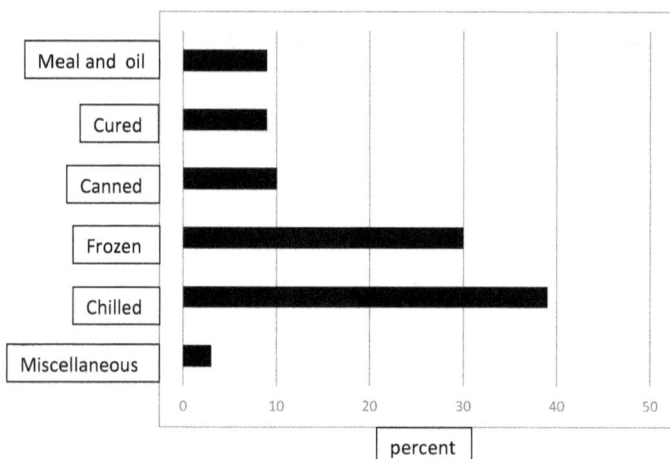

Figure 1.4 *World aquaculture and commercial catches, by area, 2017.*

Source: Food and Agriculture Organization of the United Nations (2020)

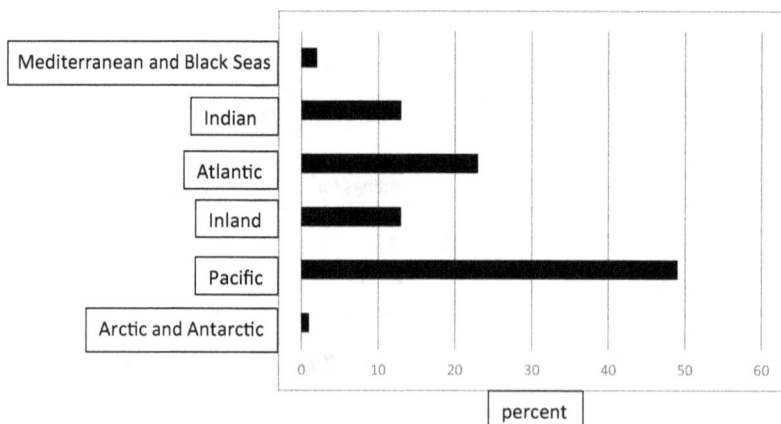

Figure 1.5 *Disposition of world aquaculture and commercial catches, 2017.*

Source: Food and Agriculture Organization of the United Nations (2020)

Table 1.1 Examples of Processing Recovery of Selected Commercially Harvested North Pacific Fishes; Expressed as Percent of Yield from Round (Whole, Unprocessed) Fish to Skinless Fillets

Fish	Species	Recovery %	Inferred Discard %
Pacific cod	*Gadus macrocephalus*	33	67
Pacific halibut	*Hippoglossus stenolepis*	41	59
Pacific herring	*Clupea pallasii*	49	51
Sockeye salmon	*Oncorhynchus nerka*	46	54
Yellowfin sole	*Limanda aspera*	25	75

Source: Crapo et al. (2004)

As much as 70% of the fisheries harvests which are used in industrial processing, and thus enter the seafood supply chain, are composed of "byproducts", which are often reduced to low-value meal and oil products, or, all too often, simply discarded (Table 1.1). For example, less than 50% of several abundant Alaska fishes is recovered during processing; the rest is discarded (= byproducts). These byproducts—the unused parts of a finfish, including viscera, heads, skin, bones, and fins—are rich sources of marine nutraceuticals and thus are the focus of this chapter.

1.2 Opportunity

The nutraceutical market offers significant opportunity to seafood producers. The global nutraceutical market size was valued at USD 417.66 billion in 2020 and is expected to expand at a compound annual growth rate (CAGR) of 8.9%, to reach USD 826.43 billion by 2028 (Grandview Research 2021).

The inter-related trends that drive this amazing growth include—

1.2.1 Rise in Chronic and Non-Communicable Disorders

The World Health Organization reports that, in 2019, seven of the top ten global causes of death were chronic and non-communicable—

Rank

1 Heart disease
2 Stroke
3 Chronic obstructive pulmonary disease
6 Trachea, bronchus, lung cancers
7 Alzheimer disease and related dementia
9 Diabetes mellitus
10 Kidney diseases

And to this list can be added ailments such as hypertension, allergies, vascular disorders, gastrointestinal conditions, endocrine disorders, and chronic (non-healing) musculoskeletal injuries and deterioration.

Well-established among developed countries, these disorders now appear to be affecting people in the developing economies, as urbanization and globalization accelerate.

1.2.2 Increasing Health Care Costs

This well-known trend shows no sign of slackening and is becoming much worse as the COVID-19 pandemic drags on. Spending on health care is reportedly growing faster than the overall economy.

1.2.3 Major Demographic Segments

As they age, Baby Boomers are increasingly interested in maintaining health and enjoyment of life well into retirement. They have the available funds to spend on a wide variety of nutraceuticals and functional foods. Further, Millennials are earning more, which allows them to afford many nutraceuticals and nutritional supplements to support and improve their active lifestyles.

1.2.4 Desire for Natural Products and Remedies

Natural products and remedies, nutritional supplements and functional foods, and similar products are continuing their expansion. In part, this is driven by concern and confusion about the side effects of prescription medications, as well as their cost and availability during the COVID-19 pandemic.

1.2.5 Parallel Trends

Along with the growth in nutraceuticals, there is similar growth in the sectors of weight management and wellness-focused diets.

1.2.6 Do-It-Yourself Health Care

The COVID-19 pandemic has ignited, or perhaps accelerated, a trend toward consumers taking some health matters and questions into their own hands. The frightening over-crowding of hospitals, the unconscionable stress on health care providers, and the risks of going anywhere and doing anything certainly contribute to this. This trend parallels the explosive interest in at-home fitness equipment.

1.2.7 Convenience and Availability

Last, but not least, is the desire to achieve, maintain, and improve one's health without too much difficulty. Information is readily accessible via the internet, as is availability of nutraceuticals from myriad online e-tailers.

The global nutraceutical market size was valued at USD 382.51 billion in 2019 and is expected to expand at a compound annual growth rate (CAGR) of 8.3%, to reach USD 722.49 billion by 2027 (Grandview Research 2022). Proteins, amino acids, and omega-3 fatty acids are expected to grow substantially.

* * * * * * * * * * * * * * * *

The next several sections of this chapter draw from several recent reviews, which are cited at the end, in a "mix-and-match" fashion (Ashraf et al 2020; Atef and Ojagh 2017; Caruso et al 2020; Shahidi et al 2019; and Simat et al 2020). This approach is useful to a seafood processor because—

- There is much overlap among the reviews
- Although each review says that certain nutraceuticals can be obtained from specific species, many such substances can be obtained from pretty much any species of finfish
  - However there are some species-specific quantitative differences; for example, fish oil (as a byproduct) is usually obtained from the livers of cod (*Gadus* spp) vs. the heads of salmon (*Oncorhynchus* spp)

1.3 Omega-3 Polyunsaturated Fatty Acids

Omega-3 fatty acids are the most well-known and well-studied fish-based nutraceuticals. A fatty acid is a chain of carbon atoms with an organic acid group (COOH) at one end and an omega or methyl group (CH_3) at the other end. They are categorized based on various characteristics, such as length, the existence of double bonds, and the arrangement of hydrogen atoms in double bonds.

There are mainly three types of fatty acids—

- Saturated fatty acids (SFAs)
- Monounsaturated fatty acids (MUFAs)
- Polyunsaturated fatty acids (PUFAs)

Humans are able to synthesize SFAs and MUFAs endogenously, but we cannot synthesize PUFAs. Therefore, we must obtain these essential nutrients (termed "essential fatty acids") from our diet. They help in the formation of healthy cell membranes, proper development and functioning of the brain and nervous system, and production of hormone-like substances called eicosanoid, thromboxane, leukotriene, and prostaglandin. They are also responsible for regulating blood pressure, blood viscosity, vasoconstriction, and immune and inflammatory responses.

There are two types of PUFAs, which are classified as omega-3 (ω−3) and omega-6 (ω−6) based on the location of the last double bond relative to the terminal methyl end of the molecule. Of these, ω–3 and ω–6 PUFAs play the most important biological roles.

The balance between ω–3 and ω–6 PUFAs is believed to be a crucial factor in many disease states including cardiovascular diseases. It is important to strike a balance between the two nutrients. Today, thanks to technological advances and modern farming practices, many people in developed economies eat far more ω–6 fatty acids than ω–3.

That's a problem because while ω–3 are anti-inflammatory, ω–6 tend to be pro-inflammatory. So when ω–6 intake is high and ω–3 intake is low, the result is excess inflammation and boost in the production of body fat. The large ω–6:ω–3 imbalance in the Western diet has been tied to more than just obesity. It's also been linked to diabetes, heart disease, certain cancers, depression, pain, inflammatory conditions like asthma, and autoimmune illnesses.

The major ω–3 fatty acids are alpha-linolenic acid (ALA), eicosapentaenoic acid (EPA), and docosahexaenoic acid (DHA). ALA is a precursor of EPA and DHA. EPA and DHA are found mainly in fatty fishes such as mackerel, sardines, anchovies, Pacific salmon, herring, trout, tuna, and in fish oils. These fatty acids are known to be essential in the growth of children and prevention of coronary heart diseases. DHA is important for optimal brain and neurodevelopment in children, and EPA helps in improving cardiovascular health overall. Fatty acids also play an important role in membrane mediated processes such as osmoregulation, nutrient assimilation, and transport.

The health benefits of ω–3 fatty acids are well-known to scientific, clinical, and industry experts, with research examining effects on almost every body system and for numerous health conditions (Health.com 2022 and Healthline. com 2022). Omega-3 fatty acids offer myriad health benefits promoting health and healing in our major organ systems—

1.3.1 Cardiovascular System

A 2011 evaluation of 17 studies indicated that people who eat seafood (fish and shellfish) one to four times a week are less likely to die of heart disease than those who rarely or never eat seafood.

- Eating seafood (fish and shellfish) has been linked to a moderate reduction in the risk of stroke.
- There is some evidence that ω–3s from marine sources (such as fish oil) may reduce the risk of one type of stroke (ischemic stroke—the type caused by narrowing or blockage of a blood vessel in the brain), but ω–3s have not been shown to reduce total strokes or death from stroke.
- Triglycerides are a type of fat found in people's blood. Excessive levels of triglycerides may raise the risk of heart disease. Dietary changes, weight control, and exercise are used to lower triglyceride levels. Some people also need to take medicine to lower their triglyceride levels.

The World Health Organization reports that heart disease is the leading cause of death worldwide. Studies show that people who eat a lot of fish have much lower rates of heart disease.

Multiple risk factors for heart disease appear to be reduced by consumption of fish or fish oil. The benefits of fish oil for heart health include:

- **Cholesterol levels:** It can increase levels of "good" cholesterol (HDL, high-density lipoprotein). However, it does not appear to reduce levels of "bad" cholesterol (LDL, low-density lipoprotein).
- **Triglycerides:** It can lower triglycerides ("blood fats").
- **Blood pressure:** Even in small doses, it helps reduce blood pressure in people with elevated levels.
- **Plaque:** It may prevent the plaques that cause your arteries to harden, as well as make arterial plaques more stable and safer in those who already have them.
- **Fatal arrhythmias:** In people who are at risk, it may reduce fatal arrhythmia events. Arrhythmias are abnormal heart rhythms that can cause heart attacks in certain cases.

Although fish oil supplements can improve many of the risk factors for heart disease, there is no clear evidence that it can prevent heart attacks or strokes. Therefore, experts believe fish oil may support the health of your heart.

1.3.2 Brain and Nervous System

The human brain is made up of nearly 60% fat, and much of this fat is $\omega-3$ fatty acids. That is why $\omega-3$s are essential for normal brain function. In one study, fish oil improved cognitive performance in healthy adults between the ages of 51 and 72 in just five weeks, compared with the effects of a placebo.

As we age, our brain function slows down, and the risk of Alzheimer's disease and related dementia increases. People who eat more fish have slower age-related mental decline. However, it's unclear if fish oil supplements can prevent or improve mental decline in older adults.

Depression is forecast to become the second-largest cause of illness by 2030 (Mathers & Loncar 2006). People with major depression appear to have lower blood levels of $\omega-3$s. Studies show that fish oil and $\omega-3$ supplements may improve symptoms of depression. Other research has also connected higher blood levels of $\omega-3$s with a lower risk of depression and anxiety. When used as an adjunct to standard antidepressant therapies, fish oil supplements are beneficial in the treatment of depression compared to a placebo. Also, some studies have shown that oils rich in EPA are especially helpful in relieving depressive symptoms.

Interestingly, some studies indicate that people with certain mental disorders have lower ω–3 blood levels, and that fish oil supplements can prevent the onset or improve the symptoms of some mental disorders. For example, it can reduce the chances and symptoms of psychotic disorders in those who are at risk of both schizophrenia and bipolar disorder.

Finally, behavioral disorders in children can interfere with learning and development, and fish oil supplements have been shown to help reduce hyperactivity, inattention, impulsiveness, aggression, and other negative behaviors.

1.3.3 Vision System

Omega-3 fatty acids are also important to the eyes, and a low intake of ω–3 fatty acids is linked to a greater risk of eye diseases.

Further our eyes begin to decline as we age, and this can lead to age-related macular degeneration (AMD). Eating fish is linked to a reduced risk of AMD, but the results on fish oil supplements are less convincing.

Omega-3 fatty acids are also thought to reduce dry eye disease and retinitis pigmentosa.

1.3.4 Musculoskeletal System

In the typical American diet, it's common to consume far more ω–6 fatty acids—which are found in plant oils, like corn and sunflower oils—than ω–3 fatty acids, particularly DHA and EPA. During old age, bones can begin to lose their essential minerals, making them weaker and more likely to break. This can lead to conditions like osteoporosis and osteoarthritis.

The ω–6:ω–3 imbalance has been linked to low bone density, but older adults with higher ω–3 levels appear to maintain greater bone density, making fish oil a potential mediator of age-related bone loss.

1.3.5 Anti-Inflammation

Inflammation is how the immune system fights infection and treats injuries. Chronic inflammation is associated with serious disorders, such as obesity, diabetes, depression, and heart disease, but reducing inflammation can help treat symptoms of these disorders.

Because fish oil has strong anti-inflammatory properties, it appears to help treat causes and symptoms of chronic inflammation, such as joint pain and stiffness, including the pain of rheumatoid arthritis.

1.3.6 Skin

The skin is the largest organ in the body, and it is rich in ω–3 fatty acids. Our skin health declines throughout our life, especially as we age or experience too much sun exposure. Fish oil is thought to reduce skin disorders such as dermatitis and psoriasis, which maintains skin health.

1.3.7 Other Benefits

The ω–3 fatty acids in fish and fish oil are reported to support—

- Pregnancy and early infant development
- Liver function
- Lung function, especially reduction of childhood asthma and allergies
- Weight loss and general body condition

The global fish oil market size was valued at $1,905.77 million in 2019, and is estimated to reach $2,844.12 million by 2027 with a CAGR of 5.79% from 2021 to 2027 (Allied Market Research 2020).

1.4 Proteins and Peptides

People have eaten fish since they were able to catch them, and the benefits of eating fish (especially as alternative to mammal or bird protein) have been widely known for a long time.

Proteins are composed of peptides, which are composed of amino acids. Proteins are large, complex, and versatile compounds, which offer an extremely wide range of nutraceutical benefits. This section will outline the benefits which can be obtained from four general types of products (Figure 1.6).

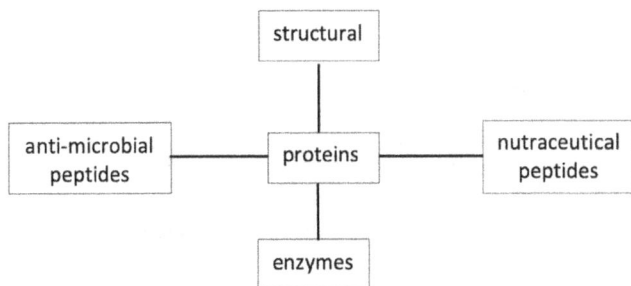

Figure 1.6 *Fish-based proteins and peptides.*

1.4.1 Structural Proteins

Structural proteins are not considered to have much biochemical activity (as enzymes do), and so are not strictly included as nutraceuticals. But they are extremely useful, and increasingly common in the nutraceutical sector.

Collagen composes about 25% of all protein in vertebrates (it's the primary structural protein in Animalia). Abundant in glycine, proline, and hydroxyproline, it has many uses, based on its biocompatibility, its ability to form strong fibers, and its stability by formation of cross-links. Of the many types of collagen, the most useful to the nutraceutical industry are—

• Type 1: found in fish skin (and in bone, mineralized with hydroxyapatite)
• Type 2: found in fish cartilage

Gelatin is a partially hydrolyzed form of collagen, with many uses in the pharmaceutical, nutraceutical, cosmeceutical, and food processing industries. It is made by denaturing collagen with dilute acids in heated solutions.

Collagen and gelatin can be obtained from any vertebrate species, but fishes offer several advantages over mammalian or avian sources. Collagen and gelatin derived from mammals have a number of shortcomings, including immunogenicity, batch-to-batch variation, and pathogenic contamination. Further, religious and cultural beliefs can conflict with and limit treatment options, especially in surgery. Jewish, Muslim, and Hindu people are generally prohibited from consuming porcine and bovine products, in foods and in medical applications.

In contrast, fish collagen and gelatin offer—

• Abundant availability
• Low immunogenicity
• Very low risk of disease transmission
• No basis for moral or religious objection

Fish-based collagen and gelatin are used in cosmetics and cosmeceuticals, and as edible coatings on meats, poultry, seafood, and plant-based meat substitutes.

Collagen is often used in tissue engineering: wound healing (dermal filler and scaffold for regrowth), remedy for osteoarthritis pain, and intra-arterial stents and other implants. Gelatin is used in drug delivery, such as microencapsulation of vitamins and pharmaceuticals.

1.4.2 Enzymes, Nutraceutical Peptides, and Amino Acids

Enzymes are proteins which are biochemically active (often termed "functional proteins" to distinguish them from structural proteins), and the peptides of interest in this chapter also have biochemical activity. This is an extremely broad and diverse group of compounds, which present a fascinating array of benefits to human health, based on their abundance of essential amino acids (Nguyen et al 2020; Pavlicevic et al 2020; and Shukla 2016)—

- Regulation of important metabolic pathways
- Anti-oxidant effects
- Importance for growth and development of the human body
- Angiotensin converting enzyme inhibitors (ACE inhibitors)
- Anti-hypertensive effects
- Maintenance and repair of tissues
- Anti-diabetic/lower glycemic index/lower insulin resistance
- Maintenance of electrolyte balance
- Anti-cancer/anti-proliferative
- Reduction of metabolic disorders
- Anti-coagulant
- Anti-cholesterol
- Calcium binding
- Immunomodulation
- Increased leptin/anti-hunger

Amino acids are the building blocks of peptides and enzymes. Amino acids are small organic molecules composed of an amino group ($-NH_{+3}$), a carboxylate group ($-CO_{-2}$), and a side chain which is specific for each amino acid. An essential amino acid (EAA) cannot be synthesized by humans, and must be obtained from the diet; they include: phenylalanine, histidine, lysine, valine, leucine, isoleucine, methionine, threonine, and tryptophan.

On the other end of the molecular size spectrum, enzymes are very large molecules composed of a great many amino acids—hundreds and thousands. Peptides are chains of amino acids, much smaller than proteins. Peptides are obtained by cleaving proteins into selected sections.

The gastrointestinal tract, in particular, is known to be a rich source of enzymes and peptides. This large and extremely complex group of fish-based nutraceuticals offers both exciting opportunity (many uses and applications in a growing market) and significant challenge (complex methods of extraction, isolation, and purification, in a high-volume commercial processing plant) to a seafood producer.

1.5 Carotenoid Pigments

These red, orange, and yellow pigments provide a wide range of colors to finfishes, and are generally associated with the quality of the products. In the aquatic environment, they are mostly produced by microalgae and macroalgae, and are passed and accumulated up the food chain into fishes. The carotenoids (also called tetraterpenoids) most commonly found in fishes include:

lutein, alpha carotene, beta carotene, zeaxanthin, canthaxanthin, and astaxanthin. Astaxanthin is added to feeds for aquacultured salmon, to impart the desirable red color to the flesh.

Carotenoids benefit human health because they reduce the oxidative stress linked to various reactive oxygen species (ROS) related disorders, including various types of cancer, neurological and cardiovascular diseases. They also—

- Protect eye macula from being damaged by blue light
- Protect low-density lipoprotein ("good cholesterol") against oxidation
- Reduce skin inflammation
- In topical gels and creams, provide sun protection

1.6 Chondroitin, Glucosamine, and Hyaluronic Acid

Chondroitin is an important structural component of cartilage and helps it to resist compression. Glucosamine is a crystalline compound which occurs widely in connective tissue. There are several commercially available products (dietary supplements) which provide combinations of chondroitin and glucosamine as very effective treatment of osteoarthritis and for reducing arthritis pain. Glucosamine also offers anti-inflammatory benefits.

Fish-based hyaluronic acid has a wide range of applications in biomedical applications such as tissue engineering, drug delivery, and certain surgical procedures.

1.7 Other Nutraceuticals

This chapter has presented the fish-based nutraceuticals of greatest potential interest to commercial seafood producers, based on—

- The abundance of the major capture- and culture-origin species groups
- The copious and accessible scientific foundation for exploring these opportunities
- The apparent markets for the popular products

But it should be noted that there are several other, lesser-known nutraceuticals of fish origin, which include—

- Essential amino acids
- Lipids (fats)
- Probiotics
- Vitamins
- Minerals

The interested reader is invited to explore these avenues after gaining a solid footing in producing popular products for well-known markets and market channels. (Please see the "cautionary valediction" at the end of this chapter.)

1.8 Possible Contaminants

Seafood producers who are exploring opportunities in fish-based nutraceuticals should be aware of possible contaminants such as methyl mercury, polychlorinated biphenyls (PCBs), several types of organochlorine herbicides and pesticides, per- and poly-fluoroalkyl substances (PFASs), microplastics, hormones, and other environmentally acquired chemicals.

Although these chemicals are, alas, ubiquitous (albeit in very low levels, usually), the process of extracting and concentrating beneficial nutraceuticals might also concentrate the harmful contaminants, as well.

The prudent seafood producer will—

- Select fish species which are known to be low in contaminants
- Test the nutraceutical products to confirm the absence of contaminants

In general, fish species which are not long-lived and not near the top of the marine food pyramid are safer bets, because they are less likely to have bioaccumulated harmful concentrations of contaminants. Salmon (*Oncorhynchus* spp), pollock and cod (*Gadus* spp), sardine and pilchard (*Sardina* spp, *Sardinella* spp), herring (*Clupea* spp), and anchovy (*Engraulis* spp) are popular examples of this category. Information on contaminants in most commercial species is available from many government, university, and trade association sources.

1.9 Next Steps for Seafood Producers

Every seafood producer seeks to minimize waste (for both environmental and economic reasons) and to maximize profit by maximizing the value of the products. This presents a strong incentive to explore opportunities in fish-based nutraceuticals.

To get started, download the review papers (cited at the end of this chapter) and determine if the species that you process are listed as the known sources of nutraceuticals. If not, look for related species.

Then request assistance from experts at universities, government agencies, trade associations, and possibly non-governmental organizations. For example, in the US, many universities support the Sea Grant Marine Advisory Program, which is a rich source of information and expert guidance. In Alaska, seafood producers can also call upon—

- Alaska Seafood Marketing Institute
- Alaska Fisheries Development Foundation

From there, the process is fairly straightforward. Begin with an inventory of the species that you produce or process (Figure 1.7), either seasonally or year-round.

Second, determine what parts of your waste stream are (1) most readily available and capable of separation from other parts (skin vs. guts, frames vs. heads, etc.), and (2) most problematic (messy, polluting, expensive, etc.).

Third, figure out what products can be produced from those candidate waste streams, such as oil from heads or livers, or collagen from skin.

Fourth, examine the possible markets for those products. This can be daunting, but readily possible via the internet (Amazon, Ali Baba, and many search engines).

Fifth, go back to the cited review papers (perhaps with assistance from your local experts) to outline the methods of extraction or production that best suits your operations.

Sixth, take a look at the competition—Are there many competing producers of your intended products? What makes your product better (price, quality, features, etc.)? Also, how do the competitors distribute and sell their products—directly online, through wholesalers, or via distributors in large markets?

Seventh, now start putting the puzzle together. Draw your own flow chart (Gantt, PERT, or simple box-and-arrow) for your production flow. Be as quantitative as possible—estimate the quantity, time, and cost for each step.

Finally, compare your forecasts of expected revenue with expected costs, and evaluate the potential profitability of your business model. Be very conservative—deliberately underestimate the revenue and deliberately overestimate the costs.

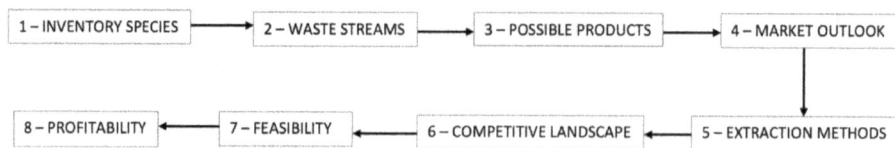

```
1 – INVENTORY SPECIES  →  2 – WASTE STREAMS  →  3 – POSSIBLE PRODUCTS  →  4 – MARKET OUTLOOK
                                                                                    ↓
8 – PROFITABILITY  ←  7 – FEASIBILITY  ←  6 – COMPETITIVE LANDSCAPE  ←  5 – EXTRACTION METHODS
```

Figure 1.7 *Flow chart of steps to determine what fish-based nutraceuticals are most appropriate for your seafood operations.*

A cautionary valediction—do not try to do everything all at once. Select the one or two most likely opportunities ("low-hanging fruit") for your first venture.

* * * * * * * * * * * * * * * *

Reviews

Ashraf, SA, M Adnan, M Patel, AJ Siddiqui, M Sachidanandan, M Snoussi, & S Hadi. 2020. Fish-Based Bioactives as Potent Nutraceuticals: Exploring the Therapeutic Perspective of Sustainable Food from the Sea. *Marine Drugs* 18:265. http://doi.org/10.3390/md18050265

Atef, M, & SM Ojagh. 2017. Health Benefits and Food Applications of Bioactive Compounds from Fish Byproducts: A Review. *Journal of Functional Foods* 35(2017):673–681. http://doi.org/10.1016/j.jff.2017.06.034

Caruso, G, R Floris, C Serangeli, & L Di Paola. 2020. Fishery Wastes as a Yet Undiscovered Treasure from the Sea: Biomolecules Sources, Extraction Methods and Valorization. *Marine Drugs* 18:622. http://doi.org/10.3390/md18120622

Shahidi, F, V Varatharajan, H Peng, & R Senadheera. 2019. Utilization of Marine By-products for the Recovery of Value-added Products. *Journal of Food Bioactives* 6:10–61.

Simat, V, N Elabed, P Kulawik, Z Ceylan, E Jamroz, H Yazgan, M Cagalj, JM Regenstein, & F Ozogul. 2020. Recent Advances in Marine-Based Nutraceuticals and Their Health Benefits. *Marine Drugs* 18:627. http://doi.org/10.3390/md18120627

* * * * * * * * * * * * * * * *

More Information and Statistics

Allied Market Research. 2020. *Fish Oil Market by Species (Anchovy, Mackerel, Sardines, Cod, Herring, Menhaden and Others), and (Aquaculture, Animal Nutrition & Pet Food, Pharmaceuticals, Supplements & Functional Food and Others): Global Opportunity Analysis and Industry Forecast, 2021–2027.* www.alliedmarketresearch.com/fish-oil-market, accessed 15 Feb 22.

Crapo, C, B Paust, & J Babbitt. 2004. *Recoveries and Yields From Pacific Fish and Shellfish. Alaska Sea Grant College Program*, Marine Advisory Bulletin No. 37. https://seagrant.uaf.edu/bookstore/pubs/MAB-37.html

FAO. 2020. *FAO Yearbook: Fishery and Aquaculture Statistics 2018*. Rome. https://doi.org/10.4060/cb1213t

Grandview Research. 2022. *Nutraceuticals Market Size, Share & Trends Analysis Report by Product (Dietary Supplements, Functional Food, Functional Beverages), By Region (North America, Europe, APAC, CSA, MEA), and Segment Forecasts, 2021–2030.* www.grandviewresearch.com/industry-analysis/nutraceuticals-market, accessed 15 Feb 22.

Health.com. 2022. *7 Potential Benefits of Fish Oil, According to a Nutritionist*. www.health.com/nutrition/fish-oil-benefits, accessed 15 Feb 22.

Healthline.com. 2022. *13 Benefits of Taking Fish Oil*. www.healthline.com/nutrition/13-benefits-of-fish-oil, accessed 15 Feb 22.

Mathers, CD, & D Loncar. 2006 Projections of Global Mortality and Burden of Disease from 2002 to 2030. *PLoS Medicine* 3(11):e442. http://doi.org/10.1371/journal.pmed.0030442

National Marine Fisheries Service. 2021. *Fisheries of the United States, 2019*. U.S. Department of Commerce, NOAA Current Fishery Statistics No. 2019. www.fisheries.noaa.gov/national/sustainable-fisheries/fisheries-united-states

Nguyen, TT, K Heimann, & W Zhang. 2020. Protein Recovery from Underutilised Marine Bioresources for Product Development with Nutraceutical and Pharmaceutical Bioactivities. *Marine Drugs* 18:391. http://doi.org/10.3390/md18080391

Pavlicevic, M, E Maestri, & M Marmiroli. 2020. Marine Bioactive Peptides—An Overview of Generation, Structure, and Application with a Focus on Food Sources. *Marine Drugs* 18:424. http://doi.org/10.3390/md18080424

Shukla, S. 2016. Therapeutic Importance of Peptides from Marine Sources: A Mini Review. *Indian Journal of Geo Marine Sciences* 45:11.

2

Bioactive Components from Marine Fungi and Their Significance as Nutraceuticals, Pharmaceuticals, and Cosmeceuticals

Maya Raman

Contents

2.1 Introduction

Nutraceuticals (functional foods and dietary supplements) have remarkable market potential (Mordor Intelligence Global Nutraceuticals Market—Growth, Trends and Forecasts, 2015–2020). It was reported that the demand for these health foods was around $250 billion in 2014 and will reach $385 billion in 2020 (www.mordorintelligence.com/industry-reports/global-nutraceuticals-market-industry). The major importers of these are the USA, Europe, Japan, Asia Pacific, Middle East and Latin America (Research and Market Nutraceuticals—2012. Global Strategic Business Report Annual Estimates and Forecasts for 2010–2018, 2012). The market is dominated by the USA, Europe and Japan, accounting for 85% of the market (www.researchandmarkets.com/

DOI: 10.1201/9781003128175-2

research/n54vdx/nutraceuticals). The demand for nutraceuticals varies with different age groups. It was observed that the consumption of nutraceuticals is more common in the elderly population than in younger generations. Nutraceutical consumption is about 40% among adults in the USA; in Spain it is 90%. In the latter, the age group ranged between 35–80years and most of them were educated women (Rovira et al., 2013). The demand for nutraceuticals has caught pace in Asian countries such as India and China. These increases in demand are mainly due to greater consumer awareness toward health, increase in income level and confidence in traditional and complimentary medicines (Suleria et al., 2015).

Nutraceuticals have been in use for the last 30 years. These are natural bioactive chemical compounds that are known to promote health, prevent disease and may exhibit semi-medicinal properties. These can be obtained as a by-product from the food industry, herbal and dietary supplements, the pharmaceutical industry, bioengineered microorganisms, agro products, and/or active biomolecules. The term nutraceutical was first coined in 1989 by Stephen De Felico and he defined it as a "food, food ingredient or dietary supplement that demonstrates specific health or medical benefits including the prevention and treatment of disease beyond basic nutritional functions." Even today there is no exact definition for this term. This term originated from the nutrients and the pharmaceuticals; nevertheless, these are beyond nutrients and exhibit health/medicinal benefits. Nutraceuticals also emerged as potential cancer preventive natural sources from food.

Nature provides an abundant and wide diversity of pharmacologically active biomolecules. These natural products lead towards the development of novel pharmaceuticals, nutraceuticals/food supplements and cosmeceuticals. The marine ecosystem covers about 70% of the earth's surface and is extraordinarily rich in biological diversity, particularly in the tropical environments. According to the Global Biodiversity Assessment by the United Nations Environment Program, the ocean consists of about 178,000 marine species across 34 phyla (Mitra and Zaman, 2016); many are still unknown. Marine organisms comprise approximately half of the total biodiversity on earth and produce a wide range of novel biomolecules (Jimeno et al., 2004; Vignesh et al., 2011). In the past 50 years, exploration of marine biotopes for their unique natural products has been an important area of research. Marine microbes, specifically bacteria and fungi, have achieved considerable reputation as new promising sources of a huge number of biologically active products and bioinspired chemical drugs synthesis. Of the estimated 270,000 known natural products, 30,000compounds were reported to be obtained from marine organisms (Blunt et al., 2018). The bioactive and high-rated novel metabolites isolated from macroorganisms have already undergone clinical trials (Newman and Cragg, 2004). Therefore, attention has turned towards the less exploited resources. The researches on the chemistry and bioactivity of the vast untapped reservoir of the marine environment, specifically marine microorganisms, have increased tremendously (Bhakuni and

Rawat, 2006; Romano et al., 2017). The microorganisms have the advantage of macroorganisms in sustainable production of large quantities of secondary metabolites at reasonable cost and feasible rate (Waites et al., 2009). These are also able to adapt to the special conditions and habitat and at times, accumulate unique bioactive secondary metabolites that may not be found even in terrestrial organisms (Bhakuni and Rawat, 2006). The discovery and large-scale production of penicillin during World War II from microorganisms have been able to highlight the importance of microorganisms in the food, pharmaceutical and medical industries. The discoveries of compactin and mevinolin, cholesterol synthesis inhibitors and antibiotics such as streptomycin, gentamicin, omegamycin, etc. have also enabled the understanding of the importance of microbes in therapeutics (Araki and Konoike, 1997; Larsen et al., 2007).

The marine ecosystem is highly complex and contains approximately 10^6 bacterial cells per milliliter (Azam and Worden, 2004). Marine bacteria and fungi among these are of great interest as these are rich sources of novel bioactive components. These are found on soft-bodied marine organisms that lack structural defense mechanisms and depend on the bioactive secondary metabolites produced by them or by associated microflora. Mostly these follow a sedentary life style. The biodiversity of marine microflora and the wide variety of biotherapeutics they contain, demands a thorough study and investigation in this area. The chief disadvantage of these natural products is that most of these are synthesized in the system but released extracellularly into water, which causes their rapid dilution and loss. The promising criteria regarding these organisms is that they survive in a competitive environment of unique conditions of pH, temperature, pressure, oxygen, light, nutrients and salinity; and also, these demonstrate rapid growth, reproduction and ability to survive in the ubiquitous environments. The natural bioactive components from these have been reported to prevent/treat cancer, anemia, diarrhea, obesity, diabetes, atopic dermatitis, Crohn's disease, etc. These bioactive "biomolecules" have tremendous potential for the use as active pharmaceutical ingredients (APIs) and in food supplements to design various nutraceuticals (Dewapriya and Kim, 2014). Many bacteria and fungi exist in symbiotic microbial consortia, on the surfaces of marine plants, and internal tissues of invertebrates (Wiese and Imhoff, 2019). In the last decades, a number of secondary metabolites from marine bacteria and fungi were reported (Bernan et al., 1997; Faulkner, 2001; Blunt et al., 2009), thus reflecting the growing attention towards this group of organisms by groups from academia and industry. In the year 2007, 961 new compounds were defined from marine microorganisms (Blunt et al., 2009). Between 1989–2002, around 60% of FDA-approved drugs and pre-NDA (New Drug Application) candidates were obtained from the natural environment (Chin et al., 2006). Out of these, nine were approved as medical drugs and 13 are undergoing clinical trials (Rangel and Falkenberg, 2015). As examples from fungi, the diketopiperazine halimide (or phenylahistin) obtained from the marine fungi *Aspergillus* spp., and its synthetic analog Plinabulin (NPI

2358), are in a Phase 3 clinical trial for the treatment of non-small cell lung cancer and a Phase 2 clinical trial for Neutropenia prevention (Deshmukh et al., 2018; Agrawal et al., 2018). An underlying fact is that the bioactive components obtained from marine fungi and microbes are unique and may not be produced by their counterparts in the terrestrial environment.

2.2 Nutraceuticals

Most of the bioactive components exhibit significant biological activities and are found to act as nutraceuticals for humans and animals. The wide variety and diversity have made the chemistry of these bioactive compounds novel.

2.2.1 Classification

Nutraceuticals are classified into following:

- Isoprenoid derivatives: Terpenoids, carotenoids, saponins, tocotrienols, tocopherols, terpenes
- Phenolic compounds: Couramines, tannins, lignins, anthocynins, isoflavones, flavonones, flavanoids
- Carbohydrate derivatives: Ascorbic acid, oligosaccharides, non-starch polysaccharides
- Fatty acid and structural lipids: n-3 PUFA, CLA, MUFA, sphingolipids, lecithin
- Amino acid derivatives: Amino acids, allyl-S compounds, capsainoids, isothiocyanates, indoles, folate, choline
- Microbes: Probiotics, prebiotics
- Minerals: Ca, Zn, Cu, K, Se

The chemotherapeutic values of nutraceuticals in cancer were reported earlier. Most of their cancer prevention evidence was observed in animal studies on phytochemicals, fat, flavones, phytoesterogens, isoflavones, genestein, curcumin, capsaicin, epigallocatechin-3-gallate, gingerol, lycopene, antioxidants, vitamins and minerals. Further, food ingredients like lycopene, silbinin, shark cartilage, vitamin D were cited to decrease osteoporosis and bone pain, while green tea, selenium and vitamin E, grape seed extract, modified citrus, pectin, soy, PC-SPES were listed as prostate cancer protective food supplements.

2.3 Marine Fungi

Marine fungi are widely distributed microorganisms in the ocean; and these are in association with sediment, seawater, submerged algae, marine habitants, mangroves, etc. These are divided into two groups based on their ability to grow in marine conditions, obligate and facultative marine fungi (Borse et al., 2013). Obligate marine fungi grow fast and sporulate exclusively in marine or estuarine habitat, while facultative marine fungi adapt to a marine environment. Sometimes, it is difficult to differentiate the obligate and facultative fungi; hence, the term marine-derive fungi are used (Bugni and Ireland, 2004). These are mainly involved in the decomposition of woody and herbaceous substrates, and dead animals or their parts, in the marine environment (Hyde et al., 1998). These are also important pathogens and form a symbiotic relationship with other organisms. Extreme physical and chemical conditions in the marine environment contribute to the development of specific metabolic pathways in these (Abdel-Lateff, 2008). Marine mangroves, algae, coral reefs, etc. are rich sources of marine fungi. There has been a considerable increase in the number of new bioactive compounds reported from marine fungi over the years (Rateb and Ebel, 2011).

The pivotal role of fungi in different industries such as food processing, pharmaceuticals, medicine, etc. is very pronounced and the same is the case with marine fungi. The largest family of marine fungi is Halosphaeriaceae, while *Aspergillus*, *Pencillium* and *Candida* genera are widespread (Jones et al., 2019; Kumar et al., 2015). Marine fungi belong to the phyla Ascomycota, Bacidomycota, Chytridiomycota, Deuteromycota and Zygomycota. More than 1206 marine fungal species in 472 genera are known (Jones et al., 2019; Pang and Jones, 2017). Fungi are heterotrophic eukaryotes that degrade different solid substrates and help in recycling in the marine ecosystem. Marine fungi were investigated for the infections they caused and these are known as potent pathogens (Tresner and Hayes, 1971). Despite their pathogenicity, certain marine fungi have shown symbiotic association with other marine organisms and serve as a source of biological active compounds (Duarte et al., 2012). Therefore, scientists isolating the fungi and extracting the bioactive compounds confront the serious impediment of preserving samples (Duarte et al., 2012).

Symbiosis is the term used to indicate the mutual association between two organisms for their mutual benefit, either in terms of nutritional needs or protection. Studies have indicated that a variety of secondary metabolites are produced as a result of such symbiotic associations. These associations are potential chemical and ecological phenomena and serve as sustainable resources for several novel pharmaceutical and nutraceutical leads. Certain marine fungi have shown symbiotic association (*Haliclona simulans*, *Agaricomycotina*, *Mucoromycotina*, *Saccharomycotina* and *Pezizomycotina*) (Baker et al., 2009). The extracts from these have shown antimicrobial activities against *E. coli*, *Bacillus* spp., *S. aureus* and *Candida glabrata* (Baker et al., 2009). In another study, M-3, a fungal strain, was isolated from marine red algae, *Porphyra yezoensis*; and from this, diketopiperazine was isolated, which has exhibited antifungal activity against *Pyricularia oryzae* (Byun et al., 2003). The indole-2,3-dione, a type of isatin isolated from the butanol extract of *Pseudoalteromona issachenkonii*, was reported to have shown hemolysis and inhibition against *Candida albicans* (Kalinovskaya et al., 2004). The compounds extracted from *Phoma* spp. namely, Phomadecalins A, B, C, D and Phomapentenone A, showed potential antibacterial activity against *Bacillus subtiis* (ATCC6051) and *S. aureus* (ATCC 29213) (Goldring and Pattenden, 2004). These metabolites not only have potent biological activity, they also give an upper hand to marine fungi in adapting to extreme habitats, competing for substrates and warding off threats (Abdel-Lateff, 2008).

The bioactive natural products from fungi include alkaloids, terpenoids, peptides, polyketides, steroids and lactones that can be used as antihypertensive, antioxidative, anticoagulant and antimicrobial components in functional foods or nutraceuticals, pharmaceuticals or cosmeceuticals. The marine microorganisms have an advantage that they can be cultured and hence, offer high reproducibility and an everlasting source of these natural products (Blunt et al., 2009).

2.4 Marine Fungal Peptides and Their Biological Properties

The proteins and peptides show minimal human toxicity and less adverse effects comparable to synthetic drugs (Ucak et al., 2021). The peptide antibiotics isolated from marine organisms have been explored immensely. These are classified as synthetic peptides (non-ribosomal) and natural (ribosomal) peptides. The non-ribosomal peptides include glycopeptides, gramicidins, bacitracins and polymyxins. The natural peptides possess strongly modified structures in backbones or side chains. These form good candidates for drug design and offer great stability from enzymatic degradation and thermal denaturation. Peptides are isolated from fermented fungal biomass culture media, subjected to chromatographic fractionation to obtain the pure compounds.

The cordy heptapeptides C-E isolated from *Acremonium persicinum* showed immense potentiality against MCF-7, SF-268 and NCI-H460 cancer cell lines with IC_{50} values ranging between 2.5 and 12.1µM (Chen et al., 2012). The same species was used to isolate known penta decapeptides (efrapeptin F-G, efrapeptins Eα and H) and exhibited potent cytotoxicity against the H125 cell line. Additionally, RHM1-4, an N-methylated linear octapeptide, were also isolated from same species. These peptides also exhibited pronounced antibacterial activity against *S. epidermidis* with minimum inhibitory concentration values ranging between 0.015 and 0.049µM (Kang et al., 2015). See Figure 2.1.

Five cyclopeptides—namely cyclo (Pro-Ala), cyclo (Ile-Leu), cyclo (Leu-Pro), cyclo (Pro-Gly) and cyclo (Pro-val)—were isolated from *Ascotricha* spp. Aspergillicins A-E were isolated from *Aspergillus carneus*, which showed modest cytotoxicity. Several cyclopeptides were isolated from *A. niger*, which comprised of cyclo (L-Pro-L-Phe), cyclo (*trans*-4-hydroxy-L-Pro-L-Leu), cyclo

Cordyheptapeptide C

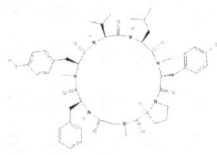

Cordyheptapeptide D

Cordyheptapeptide E

(A)

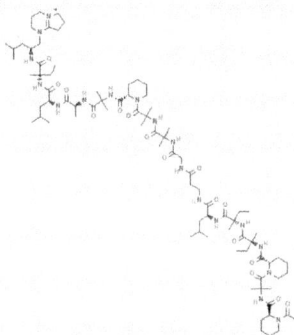

Efrapeptin H

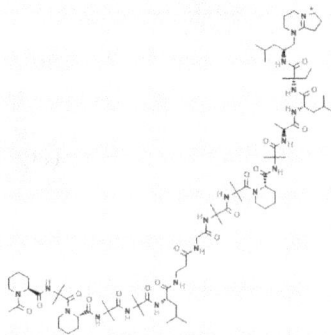

Efrapeptin F

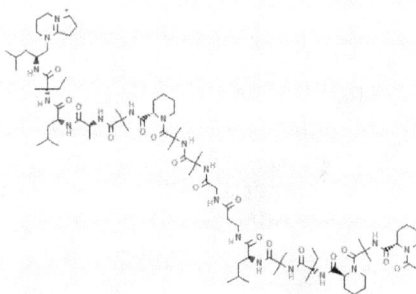

Efrapeptin G

(B)

Figure 2.1 *Structure of (A) cordy heptapeptides (B) efrapeptins.*

Source: Courtesy Youssef et al. (2019)

(L-Pro-L-Leu), cyclo (*trans*-4-hydroxy-L-Pro-L-Phe), cyclo (L-Pro-L-Val), as well as cyclo (L-Pro-L-Tyr), cyclo (L-Trp-L-Ile), cyclo (L-Trp-L-Phe) and cyclo (L-Trp-L-Tyr). Certain peptides stimulated plant growth. However, none exhibited pronounced antimicrobial activity against *E. coli*, *S. aureus* and *C. albicans* (Wang et al., 2021). The sclerotide A and B, isolated from *A. sclerotiorum* exhibited modest antifungal activity. Cyclic hexapeptide similanamide was isolated from sponge-derived *A. similanensis*, which displayed weak cytotoxicity against various cancer cell lines and no antibacterial activity. The lumazine peptides, terrelumamides A and B from *A. terreus* exhibited pronounced insulin sensitivity *in-vitro*. They exerted their action via increasing the formation of adiponectin during adipogenesis in hBM-MSCs. These also showed potent cytotoxic effects against U937 and MOLT4 human cancer cell lines. The cyclic peptides, psychrophilins E-H were isolated from *A. versicolor*, characterized by the presence of a rare linkage of amide between anthranilic acid and indole ring. The psychrophilins exhibited pronounced lipid-reducing activity at approximately 10µM. The aspersymmetide A, isolated from coral-derived *A. cersicolor*, is the first centrosymmetric cyclohexapeptide. Though, it showed weak cytotoxicity. The aspochracin-type cyclic tripeptide sclerotiotide L, diketopiperazine dimer and cyclic tetrapeptide showed potent anti-inflammatory activity against IL-10 expression of LPS-induced THP-1 cells (acute monocytic leukemia cells) at 10µM. The psychrophilin E from algae-derived *Aspergillus* spp. selectively inhibited the proliferation of colon cancer cell with IC_{50} values of 33.4µM (Ebada et al., 2014). The *Aspergillus* spp. derived peptides also exhibited antiviral activity (Figure 2.2).

2.5 Marine Fungal Secondary Metabolites and Their Biological Properties

The secondary metabolites produced by the marine fungi are mostly biologically active. These are reported to be produced by multifunctional enzyme complexes such as type-I polyketide synthases (PKS) and non-ribosomal peptide synthetases (NRPS) (Nikolouli and Mossialos, 2012). Most of the marine fungi are endophilic or epiphilic; and specific methods must be followed to isolate and cultivate (from various marine organisms such as algae, sponges and mangrove plants) in order to characterize and elucidate the structure of their secondary metabolites (Kjer et al., 2010). The isolation of bioactive secondary metabolites is still continuing. The marine fungi have been known to represent a huge reservoir of biomolecules that possess anticancer, antibacterial, antiplasmodial, anti-inflammatory and antiviral activity (Bhadury et al., 2006). This is due to the unique carbon frameworks in them. These novel compounds are used as novel lead structures for medicine and for plant protection.

These metabolites are affected by their source of isolation and the organisms they are harboring in the natural environment. The fungal metabolites have elicited an assorted biological activity including anticancer and antidiabetic

Aspergillicin A

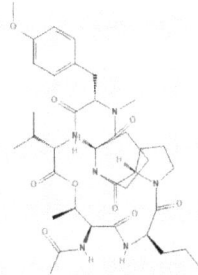

Aspergillicin B

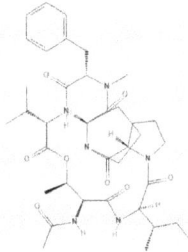

Aspergillicin C

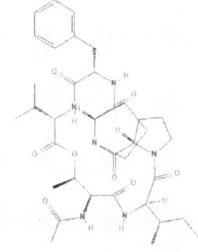

Aspergillicin D

Aspergilliin E

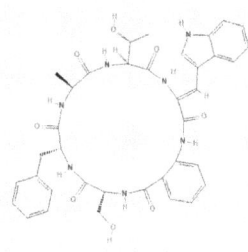

Sclerotide A

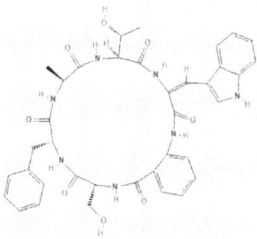

Sclerotide B

Similanamide

Figure 2.2 *Some cyclic bioactive components from* Aspergillus *spp.*

Source: Courtesy Youssef et al. (2019)

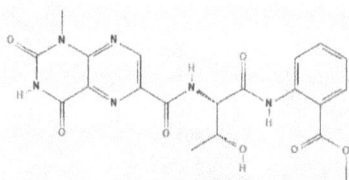

Terrelumamide A

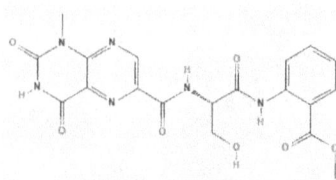

Terrelumamide B

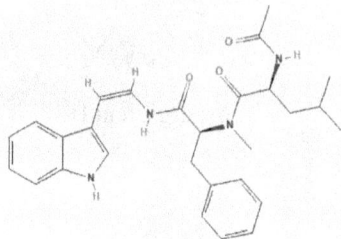

Aspergillamide A

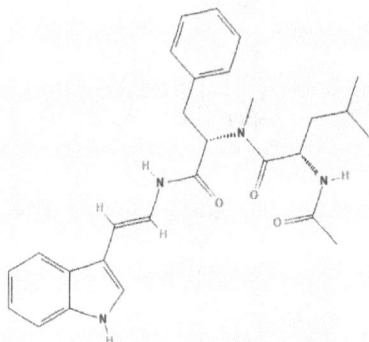

Aspergillamide B

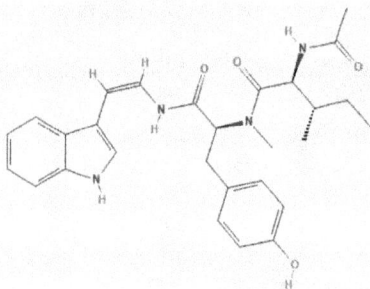

Aspergillamide C

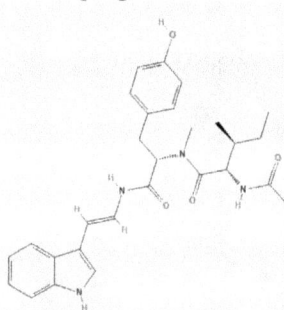

Aspergillamide D

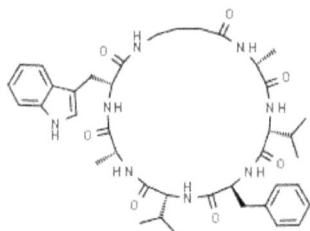

Unguisin A

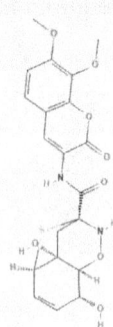

Aspergillamine A

Figure 2.2 *Continued*

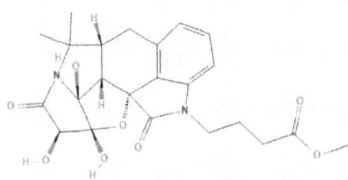

Aspergillamine B

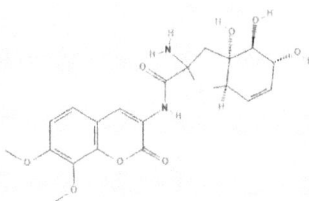

Aspergillamine C

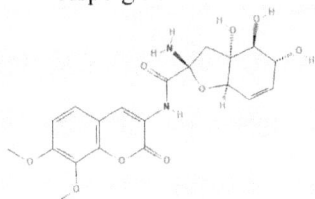

Aspergillamine D

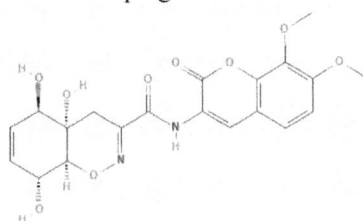

Trichodermamide A

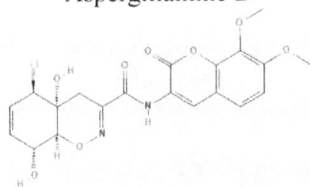

Trichodermamide B

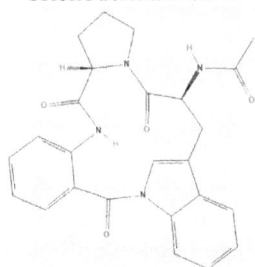

Psychrophilin E

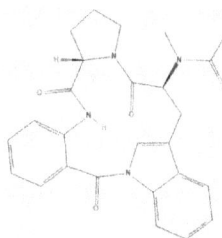

Psychrophilin F

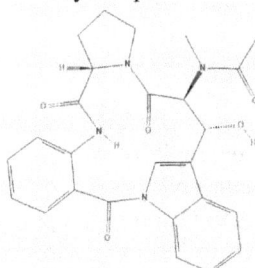

Psychrophilin G

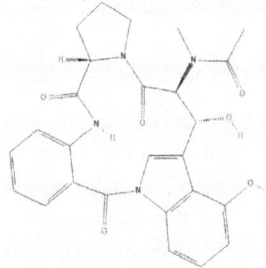

Psyhrophilin H

Figure 2.2 *Continued*

Versicotide C

Aspersymmetide A

Cotteslosin A

Cotteslosin B

Sclerotiotide L

Aspergilumamide A

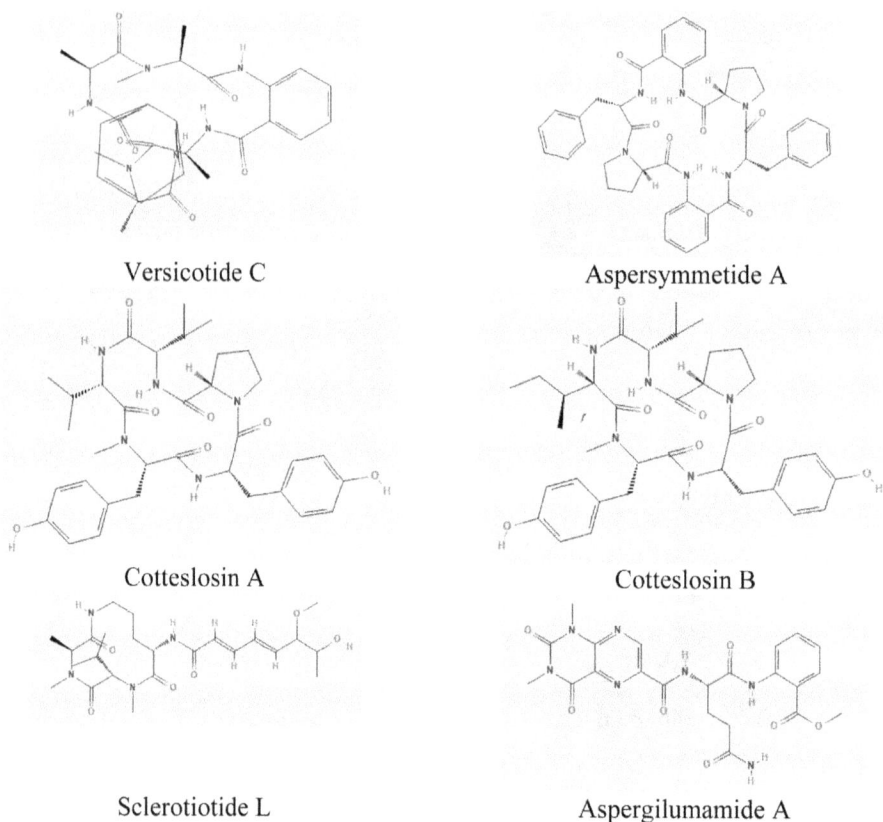

Figure 2.2 *Continued*

effects. However, the majority of them were reported to have pharmaceutical activities such as cell cycle inhibition, kinase and phosphatase inhibition, anti-oxidant, neurogenic, anti-inflammatory, antiplasmodial and antiviral activities (Agrawal et al., 2018).

2.5.1 Antibacterial, Cytotoxic, and Antiviral Effects

The marine-derived fungus *Nigrospora* spp., isolated from a sea fan (*Annella* spp.) collected near Similan Island, Thailand, is the first example of marine-derived *Nigrospora* spp., as *Nigrospora* spp. are known as plant endophytes. These produced four new metabolites termed nigrospoxydons A to C and nigrosprapyrone, together with nine known compounds, when cultured in lab in potato dextrose broth. The by-product, ethyl acetate extract, was found to have antibacterial activity against *S. aureus* and methicillin-resistant *S. aureus*. Epoxydon showed activity against both strains. The *Penicillium* spp. are known for their large variety of bioactive compounds that exhibit biological and pharmaceutical activities.

The marine-derived *Penicillium* spp. PSU-F44, isolated from sea fan, produced penicipyrone and penicilactone together with macrolides, (+) brefeldin A, (+) brefeldin C and 7-oxobrefeldin A, which showed excellent antifungal activity against *Microsporumgypseum* SH-MU-4 and antibacterial activity again *S. aureus* SK1 (Trisuwan et al., 2009). The alkaloid-rich extract from *Pencillium* spp. isolated from deep ocean sediment contained meleagrin D and E and roquefortin H and I, which showed significant antitumor activity (Kim, 2014). However, these had lower cytotoxic activity than meleagrin B and meleagrin that induced apoptosis in HL-60. The authors claimed that the distinct substitutions on the imidazole ring caused the cytotoxicity. Peniciadametizine A (1); a new dithiodiketopiperazine derivative possessing a unique spiro[furan-2,7'-pyrazino[1,2-*b*][1,2]oxazine] skeleton, together with a highly oxygenated new analogue, peniciadametizine B; as well as two known compounds, brasiliamide A and viridicatumtoxin, were isolated and identified from *Penicillium adametzioides* AS-53, a fungus obtained from an unidentified marine sponge, which showed inhibitory activity against the pathogenic *Alternaria brassicae* (Liu et al., 2015). The marine-derived fungus *Pencillium* spp. JF-55 was used to extract secondary metabolites penstyrylpyrone, anhydrofulvic acid and citromycetin. The first two compounds inhibited PTP1B (Protein tyrosine phosphatase 1B). This enzyme plays a major role in the negative regulation of insulin signaling, and is an attractive therapeutic target for preventing or treating diabetes. These also suppress the pro-inflammatory mediators via NF-κB pathway, through expression of anti-inflammatory HO-1 (hemo oxygenase-1, Lee et al., 2013). The marine *Aspergillus* spp. (Family *Trichomaceae*), isolated from the surface of *Sargassum horneri* from Gadeok Island, Korea, yielded polyoxygenated decalin derivative, dehydroxychlorofusarielin B that showed mild antibacterial activity against *S. aureus*, methicillin-resistant *S. aureus* and multidrug-resistant *S. aureus* (Nguyen et al., 2007). The anthraquinone derivative with naphtha [1, 2, 3-de]chromene-2, 7-dione skeleton was isolated from *Aspergillus glaucus* from the Fujian province of China and was termed aspergiolide A (Du et al., 2007). This exhibited cytotoxicity against leukemia cell lines (K562 and P388). Other novel compounds to be isolated and characterized were from *Ampelomyces* spp. These exhibited potent antimicrobial and antifouling activity. These also exhibited antilarvicidal activity against the larvae of tubeworm *Hydroides elegans* and of cyprids of the barnacle *Balanus amphritrite*, which is attributed to the presence of 3-chloro-2,5-dihydroxybenzyl alcohol. This compound is non-toxic and has been reported to exhibit antifouling and antibiotic activity (Kwong et al., 2006). *Cladosporium* spp. (F14 strain) also exhibited antibiotic and antifouling activity, which was attributed to the bioactive compounds produced by these by utilizing sugars such as glucose and xylose (Xiong et al., 2009). The ethanol extract from fungal strain *Fusarium* spp. (strain 05JANF165) contained Fusarielin E, which also exhibited antimitotic and antifungal activity (Hemphill et al., 2017).

Another marine-derived fungus *Pseudallescheria* spp. was reported to exhibit antibacterial activity that was attributed to novel compounds extracted from it such as dioxopiperazine, dehydroxybisdethibis-methylthio-gliotoxin. These three compounds also exhibited antibacterial activity against methicillin-resistant and multi-drug-resistant *S. aureus*. Later they exhibited significant radical scavenging activity against 1,1-diphenyl-2-picrylhydrazyl (DPPH) with IC_{50} value of 5.2μM (Li et al., 2006). The halogenated benzoquinones (bromochlorogentisylquinones A and B) extracted from the fungal strain, *Phoma hebarum*, also exhibited the radical scavenging activity (Nenkep et al., 2010). The chlorohydroaspyrones A and B, novel aspyrones from *Exophiala* spp., were reported to exhibit antibacterial activity. The other compounds formed along with these were aspyrones, asperlactone and penicillic acid. The marine ascomycete *Lachnum papyraceum* (Karst.) exhibited nematocidal effects. Many other fungi also exhibited similar activity (Stadler et al., 1995). Oxaline and Atternaramide from fungal strains have shown antitumor and antibacterial activity, respectively (Gupta and Prakash, 2019). The marine fungi isolated from sponges and gorgonian (*Cladosporium* spp.) were reported to have antibiotic activities. One new bicyclic lactam, cladosporilactam A and six known 12-membered macrolides were isolated from a gorgonian-derived *Cladosporium* sp. fungus collected from the South China Sea. Cladosporilactam A was the first example of 7-oxabicyclic [6.3.0]lactam obtained from a natural source (Bovio et al., 2019). Canakay and Yapici (2016) also reported the antifungal and antibacterial activities of marine fungi. The Scopularide A isolated from *Scopulariopsis brevicaulis* exhibited anticancer activity. The compound consists of a reduced carbon chain (3-hydroxy-methyldecanoyl) attached to five amino acids (glycine, l-valine, d-leucine, l-alanine and l-phenylalanine) (Lukassen et al., 2015). Wu et al. (2015) reported an unusual polyketide with a new carbon skeleton, lindgomycin and ascosetin from different strains of Lindgomycetaceae from the sponge of the Kiel Fjord in the Baltic Sea, Germany and from Antarctica. These compounds exhibited antibiotic activities against methicillin-resistant *S. aureus* (IC_{50} 5.1μM). Tamminen et al. (2015) discovered that agitated bioreactors are desirable for the large-scale production of calcaride A from *Calcarisporium* spp. KF525, which exhibited antibiotic activity.

Eleven new polyphenols, namely spiromastols A–K, were isolated from the deep sea-derived fungus *Spiromastix* sp. MCCC 3A00308. These are classified as diphenyl ethers, diphenyl esters, and isocoumarin derivatives; the *n*-propyl group in the analogues is rarely found in natural products. The spiromastols A-C (1-3) exhibited potent inhibitory effects against *Xanthomanes vesicatoria, Pseudomonas lachrymans, Agrobacterium tumefaciens, Ralstonia solanacearum, Bacillus thuringensis, Staphylococcus aureus* and *Bacillus subtilis*, with minimal inhibitory concentration (MIC) values ranging from 0.25–4 µg/ml (Niu et al., 2015). A new isocoumarin derivative (similanpyrone C), a new cyclohexapeptide, (similanamide) and a new pyripyropene derivative (pyripyropene T), isolated from the marine sponge-associated fungus *Aspergillus similanensis* KUFA 0013, exhibited *in-vitro* anticancer activity against MCF-7 (breast adenocarcinoma), NCI-H460 (non-small cell lung cancer) and A373 (melanoma), and antibacterial activity against multi-drug-resistant strains (Prompanya et al., 2015). Eight territrem derivatives and nine butyrolactone derivatives were isolated from marine-derived fungus *Aspergillus terreus* SCSGAF0162 under solid-state fermentation of rice. Of these, compounds 1 and 2 exhibited strong acetylcholinesterase inhibitory activity with IC_{50} values of 4.2μM. Some of these compounds also exhibited antiviral and antifouling activity (Nong et al., 2014). The marine fungus isolated from the soft coral of *Sarcophyton* spp. from the South China Sea was used to isolate a known phenylalanine derivative, a new phenylalanine derivative 4'-OMe-asperphenamate, two new cytochalasins (aspochalasin A1 and cytochalasin Z24) and eight known cytochalasin analogues. These compounds exhibited antibacterial activity against four pathogenic bacteria: *S. aureus, S. albus, E. coli and Bacillus cereus.* The cytochalasins exhibited antifouling activity against the larval settlement of barnacle *Balanus Amphitrite* (Zheng et al., 2013). The secondary metabolite, chloroazaphilones isolated from the *Bartalinia robillardoides* strain LF550 from the Mediterranean sponge *Tethya aurantium*, exhibited biological activity against bacteria, fungi, tumor cell lines and enzymes. This is the first report on the isolation of chloroazaphilones from the genus *Bartalinia* (see Table 2.1).

Table 2.1 Selected Active Compounds from Marine Fungi

Source	Compound	Activity
Nigrospora spp.	Nigrospoxydons A to C Nigrosprapyrone Epoxydon	MRSA, antibacterial
Aspergillus spp.	Dehydroxychlorofusarielin B	Antibacterial, MRSA, antiviral, cytotoxic
	Anthraquinone derivative of naphtha [1, 2, 3-de]chromene-2, 7-dione skeleton	Cytotoxic
	Similanpyrone C Similanamide Pyripyropene T	Anticancer
	Territrem derivatives Butyrolactone derivatives	Strong acetylcholinesterase inhibitory activity, antibacterial, antifouling
	11a-dehydroxyisoterreulactone A Arisugacin A Isobutyrolactone II Aspermolide A (Z)-5-(Hydroxymethyl)-2-(6')-methylhept-2-en-2'-yl)-phenol Diorcinol Cordyol C Rubrolide S Asperterrrestide A Isoaspulvinone E Aspulvinone E Pulvic acid 22-O-(N-Me-L-valyl)-21-epi-aflaquinolone B	Antiviral

(Continued)

Table 2.1 Selected Active Compounds from Marine Fungi. Continued

Source	Compound	Activity
Pencillium spp.	Penicipyrone Penicilactone Macrolides, (+) brefeldin A (+) brefeldin C 7-oxobrefeldin A	Antimicrobial, cytotoxic, antifungal
	Meleagrin D and E Roquefortin H and I	Antitumor
	Peniciadametizine A Peniciadametizine B Brasiliamide A Viridicatumtoxin	Inhibit *Alternaria brassicae* (plant pathogen)
	Penstyrylpyrone Anhydrofulvic acid Citromycetin	Antidiabetic (inhibit PTP1B)
	Purpurquinone B Purpurquinone C Purpuresters A TAN-931 Sorbicatechol A Sorbicatechol B 2-(4hydroxybenzyl)quinazoline-4(3*H*)-one 2-(4-hydroxylbenzoyl)quinazoline-4(3H)-one Methyl 4-hydroxyphenylacetate	Antiviral
Exophiala spp.	Chlorohydroaspyrones A and B Aspyrones Asperlactone Penicillic acid	Antibacterial
Alternaria spp.		MRSA
	Tetrahydroaltersolanol C Alterporriol Q	Antiviral
Ascochyta app.		Cytotoxic, antimicrobial
Cladosporium spp.	Cladosporilactam A 12-membered macrolides	Antifouling, antifungal, antimicrobial
	Oxoglycantrypine Norquinadoline A Deoxynortryptoquivaline Trytoquivaline Quinadoline B Cladosin C	Antiviral

Table 2.1 Selected Active Compounds from Marine Fungi. Continued

Source	Compound	Activity
Chaetomium spp.		Antifungal
Curvularia spp. (strain no. 768)		Cytotoxic
Spicellum roseum		Cytotoxic
Petriella spp.		Cytotoxic
Cosmospora spp.		Antidiabetic
Ampelomyces spp.	3-chloro-2,5-dihydroxybenzyl alcohol	Antimicrobial, antibiotic, antifouling, antilarvicidal
Fusarium spp. (strain 05JANF165)	Fusarielin E	Antifungal and antimitotic
Pseudallescheria spp.	Dioxopiperazine Dehydroxybisdethibis-methylthio-gliotoxin	Antibacterial, MRSA, radical scavenging
Phoma spp.	Benzoquinones (bromochlorogentisylquinones A and B)	Radical scavenging
	Phomasetin	Antiviral
Lachnum papyraceum (Karst.)		Nematocidal effect
	Oxaline Atternaramide	Antitumor, antibacterial
Scopulariopsis spp.	Scopularide A	Anticancer
Lindgomycetaceae strain	Lindgomycin Ascosetin	Antibiotic
Calcarisporium spp. KF525	Calcaride A	Antibiotic
Spiromastix sp. MCCC 3A00308	Spiromastols A–K	Antibacterial
Sarcophyton spp	4′-OMe-asperphenamate Aspochalasin A1 Cytochalasin Z24 Cytochalasin analogues	Antibacterial, antifouling,
Bartalinia robillardoides strain LF550	Chloroazaphilone	Antimicrobial, antifungal, antitumor, inhibit enzymes
Stachybotrys spp.	Stachybogrisephenone B Grisephenone A Stachyflin 3,6,8-Trihydroxyl-1-methylxanthone	Antiviral
Scytalidium spp.	Halovirus A-E	Antiviral
Ascomycetes strain 222	Balticolid	Antiviral
Fusarium spp.	Equisetin Sansalvamide A	Antiviral
Emericella spp.	Emerimidine A and B	Antiviral;
Neosartorya spp.	AGI-B4 3,4-dihydroxybenzoic acid	Antiviral

Source: Agrawal et al. (2018)

Viral diseases are one of the areas of serious disease, despite several researches, that still lack appropriate treatment and solution. The focus on the antiviral potential of marine fungi came into light in 1998, when stachyflin from *Stachybotrys* spp. RF-7260 was reported by Taishi and others. The potentiality of marine fungi in the isolation of antiviral drugs came to light in 1998, when stachyfin was isolated from *Stachybotrys* spp. RF-7260 by Minagawa and others. It exhibited promising antiviral activity against influenza A virus (H1N1) (Minagawa et al., 2002). The stachybogrisephenone B, grisephenone A and 3,6,8-Tribydroxyl-1-methylxanthone, the sesquiterpenoid and xanthone derivatives isolated from *Stachybotrys* sp. HH1ZDDS1F1-2, were reported to inhibit the *in-vitro* replication of enterovirus 71 (EV71) with the IC_{50} values of 30.1, 50 and 40.3μM (Elnaggar et al., 2016). Their inhibitory action was studied against the Coxsackie virus. Halovirs A-E, a series of lipophilic linear peptides, isolated from *Scytalidium* spp. exhibited virucidal activity against HSV-1 and HSV-2 (herpes virus) in a dose and time-dependent manner. The compound destabilized the membrane. Nong et al. (2014) reported the antiviral activity of 11a-dehyroxyisoterreulactone A, arisugacin A, isobutyrolactone II and aspernolide A derived from *Aspergillus terreus* SCSGAF0162 against HSV-1, with IC_{50} values of 33.38, 12.76, 62.08 and 68.16μM, respectively. These were isolated from gorgonian corals *Echinogorgia aurantiaca* using solid-state fermentation of rice. The equisetin and phomasetin (an enantiomeric homologue) isolated from *Fusarium heterosporum* and *Phoma* spp. exhibited *in-vitro* inhibitory activity against integrase enzyme of HIV-1, which plays a key role in replication cycle. The integric acid isolated from *Xylaria* spp., which has similar structure as equisetin, also inhibited itegrase enzyme. The stachyflin isolated from *Stachybotrys* spp. RF7260 exhibited antiviral activity against influenza A virus (H1N1). The results were comparable with zanamivir and amantadine (Minagawa et al., 2002). These suppressed the first stage of both H1N1 and H2N2 infections (Minagawa et al., 2002). The indole compounds (glyantrypine, pyrazinoquinazoline) and resultant compounds (oxoglycantrypine, norquinadoline, deoxynortryptoquivaline, deoxytryptoquivaline, tryptoquivaline and quinadoline) extracted from mangrove-derived *Cladosporium* spp. exhibited antiviral property against H1N1 strain of the influenza virus. The cladosin C extracted from the same genus from the residues of the Pacific Ocean exhibited moderate antiviral activity at 276μM. The latter was active at significantly higher concentration. The *Aspergillus sydowii* ZSDS1-F6 strain from marine sponges of the China coast contained (Z)-5-(Hydroxymenthyl)-2-(6′)-methylhept-2′-en-2′-yl)-phenol, diorcinol and cordyol C, which exhibited minor antiviral activity against H3N2 influenza virus (IC_{50} 57.4, 66.5 and 78.5μM). The rubrolides (IC_{50} 87.1μM) from *A. terreus* (OUCMDZ-1925 strain) showed potent antiviral activity against influenza H1N1 virus (Zhu et al., 2014). The gorgonian *Echinogorgia aurantiaca* from the China coast were used to isolate fungal strain *A. terreus* SCSGAF0162, which in turn was used to derive asperterrestride A (a cyclic tetrapeptide). This compound suppresses the H1N1, H3N2 strains with IC_{50} values of 18 and 8.1μM, respectively (Moghadamtousi et al., 2015). The aspulvinone, isoaspulvinone

E, aspulvinone E and pulvic acid from similar strains of mangrove soil also exhibited anti-influenza A virus activities (Moghadamtousi et al., 2015). Anti-influenza virus compounds (emerimidine A and B, emeriphenolicins A and D, aspernidine A and B and austi, austinol, dehydroaustin and acetoxydehydro-zutin), were also isolated from *Aegiceras corniculatum*. These effects were not as pronounced as others. The purpurquinones B and C, purpuresters A, TAN-931 isolated from *Pencillium purpurogenum* and sorbicatechols A and B, from *P. chrysogenum* PJX-17 exhibited anti-H1N1 activity with IC_{50} values of 85 and 113µM (Moghadamtousi et al., 2015).

Moghadamtousi et al. (2015) isolated two antiviral compounds—namely, tet-rahydroaltersolanol C and alterporriol Q—against porcine reproductive and respiratory syndrome virus (PRRSV). These compounds were isolated from marine fungus *Alternaria* spp. obtained from *Sarcophyton* spp. of China Sea. The antiviral effects of sansalvamide A, isolated from *Fusarium* spp., with IC_{50} values of 124µM were reported against molluscum contagiosum virus (MCV). Since there is a close association between MCV lesions and AIDS patients, this compound is currently being pursued for the development of new genera-tion drugs and anti-MCV agents (Moghadamtousi et al., 2015). The compound 22-O-(N-Me-L-valyl)-21-epi-aflaquinolone B isolated from *Aspergillus* spp. XS-20090B15 derived from gorgonian *Muricella abnormaliz* exhibited excep-tional anti-RSV activity (inhibitory activity against respiratory syncytial virus) with an IC_{50} value of 42 nM, roughly 500 times more potent than ribavirin. Another compound, aflaquinolone D, from the same fungus showed anti-RSV activity but not as potent as the former (IC_{50} = 6.6 µM). Similarly, 2-(4-hydroxy-benzyl) quinazolin-4(3H)-one and methyl 4-hydroxyphenylacetate, isolated from *Pencillium oxalium* strain $0312F_1$ possessed strong suppressive activity against tobacco mosaic virus with EC_{50} values of 399.57 and 829.15µM, respec-tively (Moghadamtousi et al., 2015).

2.5.2 Marine Fungi as Cosmeceuticals

Marine fungi are a good source of photo-protective compounds (mycospo-rins, mycosporine-like amino acids), caroteinoids and scytonemin (Agrawal et al., 2018). *Phaeotheca triangularis, Trimmatostroma salinum, Hortaea wer-neckii, Aureobasidium pullulans* and *Cryptococcus liquefaciens* are known to produce mycosporine (mycosporine–glutaminol–glucoside and mycospo-rine–glutamicol–glucoside), which absorb UV in the range of 310–320 nm. Similarly, marine fungi belonging to the genera *Rhodotorula, Phaffia* and *Xanthophyllomyces* are a potential source of carotenoids; which exhibit sig-nificant antioxidant and anti-inflammatory effects contributing to skin photo-protection. The marine fungus of the genus *Exophiala* was used to isolate benzodiazepine alkaloids, circumdatin I, C and G, which are more efficient than oxybenzone (ED50,350 µM) and have shown high UVA screening activ-ity with ED_{50} values of 98,101 and 105 µM, respectively (Agrawal et al., 2018). The marine fungi, *Acremonium* spp. was reported to produce hydroquinone

Table 2.2 Tyrosinase Inhibitors from Marine Fungi

Biomolecule	Source	Habitat
Myrothenones A and B (Cyclopentenone)	*Myrothecium* spp.	*Entemorpha compressa* (green algae)
6-[(E)-Hept-1-enyl]-α-pyrone (α-Pyrone Derivative)	*Botrytis* spp.	*Hyalosiphonia caespitose* (red algae)
Homethallin-II	*T. viride* H1-7	Marine sediments
1β,5α,6α,14-tetraacetoxy-9α-benzoyloxy-7β H-eudesman-2β 11-diol and 4α,5α-diacetoxy-9α-benzoyloxy-7βH-eudesman-1β,2β, 11,14-tetraol	*Pestalotiopsis* spp. Z233	*Sargassum horneri* (brown algae)

derivatives with significant antioxidant activity (Abdel-Lateff, 2008). The *Fucus vesiculosus* derived fungi *Epicoccum* spp. produce isobenzofuranone derivative (4,5,6-trihydroxy-7-methylphthalide) that has high DPPH radical scavenging activity (Abdel-Lateff, 2008). *Aspergillus wentii* EN-48 also produces metabolites with antioxidant activity. The antioxidant exopolysaccharide EPS2 was extracted from *Keissleriella* spp. YS4108 that exhibited radical scavenging activity for superoxide radicals. The diketopiperazine alkaloid golmaenone and a related alkaloid, as well as dihydroxy isoechinulin A and echinulin, derived from *Aspergillus* spp. also exhibited scavenging activity and UVA screening activity (ED_{50}, 90 and 170 M) (Agrawal et al., 2018).

Tyrosinase inhibitors are clinically potential treatment techniques for some dermatological diseases associated with melanin synthesis like hyperpigmentation, melasma, café au lait spots and solar lentigo. These are useful in cosmetic applications such as skin lightening (Agrawal et al., 2018). Table 2.2 shows the various tyrosinase inhibitors from marine fungi.

The skin acne caused by antibiotic resistant *Propionibacterium acnes* was reported to be treated by trichodin A and trichodin B, extracted from *Trichoderma* spp. strain MF106 isolated from Greenland seas. Both also showed antibiotic activity against *S. epidermidis* with IC_{50} values of 24 and 4µM, respectively.

Conclusion

The increasing rate of communicable and non-communicable health defects including bacterial and viral infections, cancer, diabetes, etc. is becoming a challenging problem in therapy. There are numerous reports on the effects of numerous drugs in preventing these diseases. However, there are many diseases for which either assured treatments are lacking or there is a tendency of recurring or multiple side effects. Therefore, natural products from different living organisms including marine organisms could be potential candidates as nutraceutical products or supplements. Nature continues to be the most significant resource in providing bioactive components that would treat diseases and also act as nutraceuticals or supplements. There are various kinds of biomolecules that can be utilized from marine fungi. The marine-derived compounds have few side effects and are more compatible with hosts; hence, there is an increased preference for these bioactive compounds. The marine-derived components also have cosmeceutical applications. The genomic information, together with advances in science and structural elucidation, has given an insight into the untapped resources from marine fungi. Further, these will improve our knowledge and will provide a great opportunity to meet the challenges in future.

References

Abdel-Lateff, A. (2008). Chaetominedione, a new tyrosine kinase inhibitor isolated from the algicolous marine fungus Chaetomium sp. *Tetrahedron Letters, 49*(45), 6398–6400.

Agrawal, S., Adholeya, A., Barrow, C. J., & Deshmukh, S. K. (2018). Marine fungi: An untapped bioresource for future cosmeceuticals. *Phytochemistry Letters, 23*, 15–20.

Araki, Y., & Konoike, T. (1997). Enantioselective total synthesis of (+)-6-epi-mevinolin and its analogs. efficient construction of the hexahydronaphthalene moiety by high pressure-promoted intramolecular diels–alder reaction of (R, 2 Z, 8 E, 10 E)-1-[(tert-Butyldimethylsilyl) oxy]-6-methyl-2, 8, 10-dodecatrien-4-one. *The Journal of Organic Chemistry, 62*(16), 5299–5309.

Azam, F., & Worden, A. Z. (2004). Microbes, molecules, and marine ecosystems. *Science, 303*(5664), 1622–1624.

Baker, P. W., Kennedy, J., Dobson, A. D., & Marchesi, J. R. (2009). Phylogenetic diversity and antimicrobial activities of fungi associated with Haliclona simulans isolated from Irish coastal waters. *Marine Biotechnology, 11*(4), 540–547.

Bernan, V. S., Greenstein, M., & Maiese, W. M. (1997). Marine microorganisms as a source of new natural products. *Advances in Applied Microbiology, 43*, 57–90.

Bhadury, P., Mohammad, B. T., & Wright, P. C. (2006). The current status of natural products from marine fungi and their potential as anti-infective agents. *Journal of Industrial Microbiology and Biotechnology, 33*(5), 325.

Bhakuni, D. S., & Rawat, D. S. (2006). *Bioactive marine natural products.* Springer Science & Business Media.

Blunt, J. W., Carroll, A. R., Copp, B. R., Davis, R. A., Keyzers, R. A., & Prinsep, M. R. (2018). Marine natural products. *Natural Product Reports, 35*(1), 8–53.

Blunt, J. W., Copp, B. R., Hu, W. P., Munro, M. H., Northcote, P. T., & Prinsep, M. R. (2009). Marine natural products. *Natural Product Reports, 26*(2), 170–244.

Borse, B. D., Borse, K. N., Pawar, N. S., & Tuwar, A. R. (2013). Marine fungi from India-XII. A revised check list. *Indian Journal of Geo-Marine Sciences, 42*(1), 110–119.

Bovio, E., Garzoli, L., Poli, A., Luganini, A., Villa, P., Musumeci, R., . . . & Varese, G. C. (2019). Marine fungi from the sponge Grantia compressa: Biodiversity, chemodiversity, and biotechnological potential. *Marine Drugs, 17*(4), 220.

Bugni, T. S., & Ireland, C. M. (2004). Marine-derived fungi: A chemically and biologically diverse group of microorganisms. *Natural Product Reports, 21*(1), 143–163.

Byun, H. G., Zhang, H., Mochizuki, M., Adachi, K., Shizuri, Y., Lee, W. J., & Kim, S. K. (2003). Novel antifungal diketopiperazine from marine fungus. *The Journal of Antibiotics, 56*(2), 102–106.

Canakay, H. M., & Yapici, B. M. (2016). Antifungal and antibacterial activities of three marine sponges obtained from the gulf of saros in Turkey. *Annals of Biological Research, 7*, 1–6.

Chen, Z., Song, Y., Chen, Y., Huang, H., Zhang, W., & Ju, J. (2012). Cyclic heptapeptides, cordyheptapeptides C–E, from the marine-derived fungus Acremonium persicinum SCSIO 115 and their cytotoxic activities. *Journal of Natural Products*, 75(6), 1215–1219.

Chin, Y. W., Balunas, M. J., Chai, H. B., & Kinghorn, A. D. (2006). Drug discovery from natural sources. *The AAPS Journal*, 8(2), E239–E253.

Deshmukh, S. K., Prakash, V., & Ranjan, N. (2018). Marine fungi: A source of potential anti-cancer compounds. *Frontiers in Microbiology*, 8, 2536.

Dewapriya, P., & Kim, S. K. (2014). Marine microorganisms: An emerging avenue in modern nutraceuticals and functional foods. *Food Research International*, 56, 115–125.

Du, L., Zhu, T., Fang, Y., Liu, H., Gu, Q., & Zhu, W. (2007). Aspergiolide A, a novel anthra-quinone derivative with naphtho [1, 2, 3-de] chromene-2, 7-dione skeleton isolated from a marine-derived fungus Aspergillus glaucus. *Tetrahedron*, 63(5), 1085–1088.

Duarte, K., Rocha-Santos, T. A., Freitas, A. C., & Duarte, A. C. (2012). Analytical techniques for discovery of bioactive compounds from marine fungi. *TrAC Trends in Analytical Chemistry*, 34, 97–110.

Ebada, S. S., Fischer, T., Klaßen, S., Hamacher, A., Roth, Y. O., Kassack, M. U., & Roth, E. H. (2014). A new cytotoxic steroid from co-fermentation of two marine alga-derived micro-organisms. *Natural Product Research*, 28(16), 1241–1245.

Elnaggar, M. S., Ebada, S. S., Ashour, M. L., Ebrahim, W., Müller, W. E., Mándi, A., . . . & Proksch, P. (2016). Xanthones and sesquiterpene derivatives from a marine-derived fungus Scopulariopsis sp. *Tetrahedron*, 72(19), 2411–2419.

Faulkner, D. J. (2001). Marine natural products. *Natural Product Reports*, 18(1), 1R–49R.

Goldring, W. P., & Pattenden, G. (2004). Total synthesis of (±)-phomactin G, a platelet acti-vating factor antagonist from the marine fungus Phoma sp. *Organic & Biomolecular Chemistry*, 2(4), 466–473.

Gupta, C., & Prakash, D. (2019). Nutraceuticals from microbes of marine sources. In *Nutraceuticals-past, present and future* (p. 99). IntechOpen.

Hemphill, C. F. P., Sureechatchaiyan, P., Kassack, M. U., Orfali, R. S., Lin, W., Daletos, G., & Proksch, P. (2017). OSMAC approach leads to new fusarielin metabolites from Fusarium tricinctum. *The Journal of Antibiotics*, 70(6), 726–732.

Hyde, K. D., Jones, E. G., Leaño, E., Pointing, S. B., Poonyth, A. D., & Vrijmoed, L. L. (1998). Role of fungi in marine ecosystems. *Biodiversity & Conservation*, 7(9), 1147–1161.

Jimeno, J., Faircloth, G., Sousa-Faro, J. M., Scheuer, P., & Rinehart, K. (2004). New marine derived anticancer therapeutics— a journey from the sea to clinical trials. *Marine Drugs*, 2(1), 14–29.

Jones, E. G., Pang, K. L., Abdel-Wahab, M. A., Scholz, B., Hyde, K. D., Boekhout, T., . . . & Norphanphoun, C. (2019). An online resource for marine fungi. *Fungal Diversity*, 96(1), 347–433.

Kalinovskaya, N. I., Ivanova, E. P., Alexeeva, Y. V., Gorshkova, N. M., Kuznetsova, T. A., Dmitrenok, A. S., & Nicolau, D. V. (2004). Low-molecular-weight, biologically active com-pounds from marine Pseudoalteromonas species. *Current Microbiology*, 48(6), 441–446.

Kang, H. K., Seo, C. H., & Park, Y. (2015). Marine peptides and their anti-infective activities. *Marine Drugs*, 13(1), 618–654.

Kim, S. K. (Ed.). (2014). *Handbook of anticancer drugs from marine origin*. Springer.

Kjer, J., Debbab, A., Aly, A. H., & Proksch, P. (2010). Methods for isolation of marine-derived endophytic fungi and their bioactive secondary products. *Nature Protocols*, 5(3), 479–490.

Kumar, A., Henrissat, B., Arvas, M., Syed, M. F., Thieme, N., Benz, J. P., . . . & Kempken, F. (2015). De novo assembly and genome analyses of the marine-derived Scopulariopsis brevicaulis strain LF580 unravels life-style traits and anticancerous scoparide bio-synthetic gene cluster. *PLoS ONE*, 10(10), e0140398.

Kwong, T. F. N., Miao, L., Li, X., & Qian, P. Y. (2006). Novel antifouling and antimicrobial compound from a marine-derived fungus Ampelomyces sp. *Marine Biotechnology*, 8(6), 634–640.

Larsen, T. O., Lange, L., Schnorr, K., Stender, S., & Frisvad, J. C. (2007). Solistatinol, a novel phenolic compactin analogue from Penicillium solitum. *Tetrahedron Letters, 48*(7), 1261–1264.

Lee, D. S., Jang, J. H., Ko, W., Kim, K. S., Sohn, J. H., Kang, M. S., . . . & Oh, H. (2013). PTP1B inhibitory and anti-inflammatory effects of secondary metabolites isolated from the marine-derived fungus Penicillium sp. JF-55. *Marine Drugs, 11*(4), 1409–1426.

Li, X., Kim, S. K., Nam, K. W., Kang, J. S., Choi, H. D., & Son, B. W. (2006). A new antibacterial dioxopiperazine alkaloid related to gliotoxin from a marine isolate of the fungus Pseudallescheria. *The Journal of Antibiotics, 59*(4), 248–250.

Liu, Y., Mándi, A., Li, X. M., Meng, L. H., Kurtán, T., & Wang, B. G. (2015). Peniciadametizine A, a dithiodiketopiperazine with a unique spiro [furan-2, 7'-pyrazino [1, 2-b][1, 2] oxazine] skeleton, and a related analogue, Peniciadametizine B, from the marine sponge-derived fungus Penicillium adametzioides. *Marine Drugs, 13*(6), 3640–3652.

Lukassen, M. B., Saei, W., Sondergaard, T. E., Tamminen, A., Kumar, A., Kempken, F., . . . & Sørensen, J. L. (2015). Identification of the scopularide biosynthetic gene cluster in Scopulariopsis brevicaulis. *Marine Drugs, 13*(7), 4331–4343.

Minagawa, K., Kouzuki, S., Tani, H., Ishii, K., Tanimoto, T., Terui, Y., & Kamigauchi, T. (2002). Novel stachyflin derivatives from Stachybotrys sp. RF-7260 Fermentation, isolation, structure elucidation and biological activities. *The Journal of Antibiotics, 55*(3), 239–248.

Mitra, A., & Zaman, S. (2016). *Basics of marine and estuarine ecology.* Springer.

Moghadamtousi, S. Z., Nikzad, S., Kadir, H. A., Abubakar, S., & Zandi, K. (2015). Potential antiviral agents from marine fungi: An overview. *Marine Drugs, 13*(7), 4520–4538.

Mordor Intelligence Global Nutraceuticals Market—Growth, Trends and Forecasts. (2015–2020). Available from: www.mordorintelligence.com/industry-reports/global-nutraceuticals-market-industry

Nenkep, V. N., Yun, K., Li, Y., Choi, H. D., Kang, J. S., & Son, B. W. (2010). New production of haloquinones, bromochlorogentisylquinones A and B, by a halide salt from a marine isolate of the fungus *Phoma herbarum. The Journal of Antibiotics, 63*(4), 199–201.

Newman, D. J., & Cragg, G. M. (2004). Marine natural products and related compounds in clinical and advanced preclinical trials. *Journal of Natural Products, 67*(8), 1216–1238.

Nguyen, H. P., Zhang, D., Lee, U., Kang, J. S., Choi, H. D., & Son, B. W. (2007). Dehydroxychlorofusarielin B, an antibacterial polyoxygenated decalin derivative from the marine-derived fungus Aspergillus sp. *Journal of Natural Products, 70*(7), 1188–1190.

Nikolouli, K., & Mossialos, D. (2012). Bioactive compounds synthesized by non-ribosomal peptide synthetases and type-I polyketide synthases discovered through genome-mining and metagenomics. *Biotechnology Letters, 34*(8), 1393–1403.

Niu, S., Liu, D., Proksch, P., Shao, Z., & Lin, W. (2015). New polyphenols from a deep sea Spiromastix sp. Fungus, and their antibacterial activities. *Marine Drugs, 13*(4), 2526–2540.

Nong, X. H., Wang, Y. F., Zhang, X. Y., Zhou, M. P., Xu, X. Y., & Qi, S. H. (2014). Territrem and butyrolactone derivatives from a marine-derived fungus Aspergillus terreus. *Marine Drugs, 12*(12), 6113–6124.

Pang, K. L., & Jones, E. G. (2017). Recent advances in marine mycology. *Botanica Marina, 60*(4), 361–362.

Prompanya, C., Fernandes, C., Cravo, S., Pinto, M. M., Dethoup, T., Silva, A., & Kijjoa, A. (2015). A new cyclic hexapeptide and a new isocoumarin derivative from the marine sponge-associated fungus Aspergillus similanensis KUFA 0013. *Marine Drugs, 13*(3), 1432–1450.

Rangel, M., & Falkenberg, M. (2015). An overview of the marine natural products in clinical trials and on the market. *Journal of Coastal Life Medicine, 3*(6), 421–428.

Rateb, M. E., & Ebel, R. (2011). Secondary metabolites of fungi from marine habitats. *Natural Product Reports, 28*(2), 290–344.

Research and Markets: Nutraceuticals—2012. Global Strategic Business Report Annual Estimates and Forecasts for 2010–2018. (2012). Available from: www.researchandmarkets.com/research/n54vdx/nutraceuticals

Romano, G., Costantini, M., Sansone, C., Lauritano, C., Ruocco, N., & Ianora, A. (2017). Marine microorganisms as a promising and sustainable source of bioactive molecules. *Marine Environmental Research, 128*, 58–69.

Rovira, M. A., Grau, M., Castañer, O., Covas, M. I., Schröder, H., & Regicor Investigators. (2013). Dietary supplement use and health-related behaviors in a Mediterranean population. *Journal of Nutrition Education and Behavior, 45*(5), 386–391.

Stadler, M., Anke, H., & Sterner, O. (1995). Metabolites with nematicidal and antimicrobial activities from the ascomycete Lachnum papyraceum (Karst.) Karst. III. Production of novel isocoumarin derivatives, isolation, and biological activities. *The Journal of Antibiotics, 48*(3), 261–266.

Suleria, H. A. R., Osborne, S., Masci, P., & Gincobe, G. (2015). Marine-based nutraceuticals: An innovative trend in the food and supplement industries. *Marine Drugs, 13*(10), 6336–6351.

Tamminen, A., Wang, Y., & Wiebe, M. G. (2015). Production of calcaride A by Calcarisporium sp. in shaken flasks and stirred bioreactors. *Marine Drugs, 13*(7), 3992–4005.

Tresner, H. D., & Hayes, J. A. (1971). Sodium chloride tolerance of terrestrial fungi. *Applied Microbiology, 22*(2), 210–213.

Trisuwan, K., Rukachaisirikul, V., Sukpondma, Y., Phongpaichit, S., Preedanon, S., & Sakayaroj, J. (2009). Lactone derivatives from the marine-derived fungus Penicillium sp. PSU-F44. *Chemical and Pharmaceutical Bulletin, 57*(10), 1100–1102.

Ucak, I., Afreen, M., Montesano, D., Carrillo, C., Tomasevic, I., Simal-Gandara, J., & Barba, F. J. (2021). Functional and bioactive properties of peptides derived from marine side streams. *Marine Drugs, 19*(2), 71.

Vignesh, S., Raja, A., & James, R. A. (2011). Marine drugs: Implication and future studies. *International Journal of Pharmacology, 7*(1), 22–30.

Waites, M. J., Morgan, N. L., Rockey, J. S., & Higton, G. (2009). *Industrial microbiology: An introduction.* John Wiley & Sons.

Wang, C., Tang, S., & Cao, S. (2021). Antimicrobial compounds from marine fungi. *Phytochemistry Reviews, 20*(1), 85–117.

Wang, Y., Qi, X., Li, D., Zhu, T., Mo, X., & Li, J. (2014). Anticancer efficacy and absorption, distribution, metabolism, and toxicity studies of Aspergiolide A in early drug development. *Drug Design, Development and Therapy, 8*, 1965.

Wiese, J., & Imhoff, J. F. (2019). Marine bacteria and fungi as promising source for new antibiotics. *Drug Development Research, 80*(1), 24–27.

Wu, B., Wiese, J., Labes, A., Kramer, A., Schmaljohann, R., & Imhoff, J. F. (2015). Lindgomycin, an unusual antibiotic polyketide from a marine fungus of the Lindgomycetaceae. *Marine Drugs, 13*(8), 4617–4632.

Xiong, H., Qi, S., Xu, Y., Miao, L., & Qian, P. Y. (2009). Antibiotic and antifouling compound production by the marine-derived fungus Cladosporium sp. F14. *Journal of Hydro-environment Research, 2*(4), 264–270.

Youssef, F. S., Ashour, M. L., Singab, A. N. B., & Wink, M. (2019). A comprehensive review of bioactive peptides from marine fungi and their biological significance. *Marine Drugs, 17*(10), 559.

Zheng, C. J., Shao, C. L., Wu, L. Y., Chen, M., Wang, K. L., Zhao, D. L., & Wang, C. Y. (2013). Bioactive phenylalanine derivatives and cytochalasins from the soft coral-derived fungus, Aspergillus elegans. *Marine Drugs, 11*(6), 2054–2068.

Zhu, T., Chen, Z., Liu, P., Wang, Y., Xin, Z., & Zhu, W. (2014). New rubrolides from the marine-derived fungus Aspergillus terreus OUCMDZ-1925. *The Journal of Antibiotics, 67*(4), 315–318.

3

Nutraceuticals from Seaweeds

Nilesh H. Joshi, Ilza R. Mor, and Rajesh V. Chudasama

Contents

3.1 Introduction

Seaweeds, named macroalgae, are an extensive group of macroscopic marine organisms that comprise of a few thousand species (Shannon & Abu-Ghannam, 2019). According to the differences of their pigmentation, marine macroalgae are generally classified into three main classes (Figure 3.1): Chlorophyceae (green seaweeds), with the pigments of chlorophyll *a* and *b* together with various characteristic xanthophylls; Phaeophyceae (brown seaweeds), including the pigments of fucoxanthin and chlorophyll *a* and *b* as photosynthetic pigment; and Rhodophyceae (red seaweeds), possessing the pigments of phycoerythrin and phycocyanin. Because of high nutritional and pharmaceutical values, seaweeds are traditionally consumed as food (e.g., sea-vegetables) or as herbal

DOI: 10.1201/9781003128175-3

43

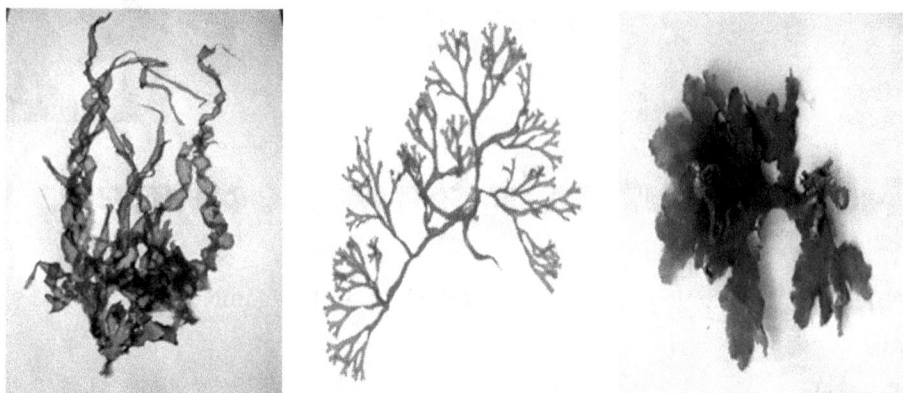

(a) Chlorophyceae   (b) Phaeophyceae   (c) Rhodophyceae

Figure 3.1 *Classes of seaweed based on pigmentation.*

medicine for treating gallstones, stomach ailments, eczema, cancer, renal disorders, scabies, psoriasis, asthma, arteriosclerosis, heart disease, lung diseases, ulcers, etc. (Lee et al., 2016; Ortiz et al., 2006).

Seaweeds are also used as fodder, fertilizer, fungicides, herbicides, condiments, dietary supplements, and as a resource of phycocolloids such as agar, alginate, and carrageenan for various industrial applications (Yaich et al., 2011). Seaweeds have become a valuable resource worldwide with a higher economic value and, consequently, an increase in basic and applied research in various related fields. The nutrients (e.g., proteins, minerals, vitamins, dietary fiber, and lipids) and numerous structurally unprecedented secondary metabolites from various species of seaweeds, including monoterpenes and meroterpenoids (Areche et al., 2009) have been reported frequently. It is noteworthy that these chemical components have demonstrated various functional properties, including nutritional and healthy functions as well as a wealth of biological activities such as anti-bacterial (Vairappan et al., 2010), antioxidant and anti-inflammatory (Chatter et al., 2009; Li et al., 2007), antiviral and anticoagulant properties (Sen et al., 1994). Moreover, these functional natural products have provided essential substances for human nutrition and promising bioactive lead compounds for drugs.

In the last three decades, interest has grown in seaweeds as nutraceuticals, or functional foods, which have provided dietary benefits beyond their macronutrient content. In addition, seaweed has been mined for metabolites with biological activity, to produce therapeutic products (Zerrifi et al., 2018). Global dietary studies have found that countries where seaweed is consumed on a regular basis have significantly less obesity and diet-related diseases (Nanri et al., 2017).

3.2 Seaweed for Protein and Amino Acids

Seaweed is rarely promoted for the nutritional value of their proteins. Protein constitutes 5% to 47% of seaweed dry mass. Among all three classed, red seaweeds have the greatest protein content, while green has less, and brown the least (Černá, 2011). Of the total amino acids in seaweeds, approximately 42% to 48% are essential (Wong & Cheung, 2000). In terms of a score (on a scale of 0.0–1.0), *Undaria pinnatifida* has an amino acid score of 1.0, equal to that of egg and soyabean, *Porphyra* 0.91, and *Laminaria saccharina* 0.82 (Murata & Nakazoe, 2001). The presence of high polyphenolic content of seaweeds can reduce the digestibility of algal proteins, giving a slightly lower score on the protein digestibility. Despite this, if other high-AA' vegan foods, such as soy or mycoprotein, are added in the diet, seaweeds can still be a viable alternative to animal-derived protein.

Seaweed proteins contain all amino acids, which significantly depend on seasonal growth (Table 3.1). According to Denis et al. (2010), the highest concentration of protein has been seen in red algae from January to April, while the lowest amount of protein was observed from July to August. The breakdown of phycobiliproteins may be linked to the lowest protein value in the summer. In contrary to the other compounds (ash or dietary fiber), proteins were put through large changes within the year. Moreover, it was reported that different protein levels depend on the specific areas, too. *Ultrica lactuca*, which comes from the littoral area of Tunisia, contains almost 50% higher protein than the same species from the Philippines (Yaich et al., 2011). Renaud and Luong-Van (2006) found that the highest concentration of protein was in red algae collected in summer (4.8%–12.8%), while it was significantly less in winter.

Many researchers have assessed crude protein content by measuring nitrogen content and multiplying it by different conversion factors. In the case of seaweed, the nitrogen-to-protein factor ranges from 3.75 to 5.72 (Černá, 2011); therefore, the traditional value might overestimate the protein content. However, nitrogen in seaweeds is a component of many types of molecules in addition to protein, such as DNA, ATP, etc.

Almost all the amino acids are presented in seaweeds (Matanjun et al., 2009). With respect to total EAAs in the FAO/WHO (1991) pattern, seaweeds (especially red and green) seem to be able to contribute to adequate levels of total EAA (Wong & Cheung, 2000). On the contrary, Matanjun et al. (2009) reported higher amounts of amino acid (AA) in green seaweeds than in red and brown.

Methionine and cysteine were detected in a high amount in red seaweeds than in green and brown, but the value showed low amounts in red algae, less than 0.3% and 0.1%, respectively (Table 3.2, Vanessa Gressler et al., 2011). The highest EAA was phenylalanine in species belonging to three groups: red, green, and brown algae (Matanjun et al., 2009). Glycine, alanine, arginine, proline, glutamic, and aspartic acids composed together a large part of the AAs fraction, whereas AAs tyrosine, methionine, and cysteine occurred in a lower amount (Gressler et al., 2010).

Table 3.1 Percentage of Protein Content in Seaweed

Seaweed Species	Protein (%)	References
Green seaweed		
Ulva fasciata	6.26 ± 0.50	Ramos et al. (2000)
Ulva compressa	21.00–32.00	Pereira (2011)
Ulva lactuca	10.00–25.00	Pereira (2011)
Ulva pertusa	20.00–26.00	Pereira (2011)
Ulva rigida	18.00–19.00	Pereira (2011)
Utricularia reticulate	17.00–20.00	Pereira (2011)
Caulerpa sertularioides	20.00 ± 1.70	Ramos et al. (2000)
Caulerpa lentillifera	10.00–13.00	Pereira (2011)
Caulerpa racemosa	17.80–18.40	Pereira (2011)
Codium fragile	8.00–11.00	Pereira (2011)
Brown seaweed		
Alaria esculenta	9.00–20.00	Pereira (2011)
Durvillaea Antarctica	10.40 ± 0.30	Ortiz et al. (2006)
Eisenia bicyclis	7.50	Pereira (2011)
Fucus spiralis	10.77	Pereira (2011)
Fucus vesiculosus	3.00–14.00	Pereira (2011)
Himanthalia elongate	5.00–15.00	Pereira (2011)
Hizikia fusiforme	11.60 ± 0.80	Ramos et al. (2000)
Laminaria digitata	8.00–15.00	Pereira (2011)
Laminaria ochroleuca	7.49	Pereira (2011)
Saccharina japonica	7.00–8.00	Pereira (2011)
Saccharina latissima	5.0–26.00	Pereira (2011)
Sargassum fusiforme	11.60	Pereira (2011)
Sargassum fluitans	12.80 ± 1.02	Ramos et al. (2000)
Sargassum polycystum	5.40 ± 0.07	Matanjun et al. (2009)
Sargassum vulgare	16.30 ± 1.30	Ramos et al. (2000)
Padina gymnospora	11.20 ± 1.00	Ramos et al. (2000)
Undaria pinnatifida	12.00–23.00	Pereira (2011)
Red seaweed		
Laurencia intricate	4.60 ± 0.0	Gressler et al. (2010)
Palmaria palmate	8.00–35.00	Pereira (2011)
Polysiphonia sp.	31.03	Dere et al. (2003)
Pyropia tenera	28.00–47.00	Pereira (2011)
Porphyra umbilivalis	29.00–39.00	Pereira (2011)
Porphyra yezoensis	31.00–44.00	Pereira (2011)
Solieria filiformis	21.25 ± 2.00	Ramos et al. (2000)
Vidalia obtusiloba	18.00 ± 1.66	Ramos et al. (2000)
Gracilaria lemaneiformis	7.76 ± 0.60	Ramos et al. (2000)
Gracilaria verrucosa	0.94	Dere et al. (2003)
Amansia multifida	25.60 ± 2.02	Ramos et al. (2000)

Source: Adapted from Pangestuti & Kim (2015)

Table 3.2 Comparison of Amino Acid Composition of Some Sea Lettuce Species (g/100 g Dry Basis)

Amino Acids	Ulva lactuca[a]	Ulva reticulata[b]	Ulva pertusa[c]	Ulva armoricana[d]
Histidine	1.8	0.23	4.00	2.10
Isoleucine	6.1	0.90	3.50	2.99
Leucine	9.2	1.68	6.90	5.22
Methionine	1.8	-	1.60	2.58
Phenylalanine	6.3	1.12	3.90	7.10
Threonine	4.6	1.15	3.10	6.88
Tryptophan	-	-	0.30	-
Valine	7.7	1.34	4.90	5.01
Lysine	6.3	1.28	4.50	4.01
Alanine	8.5	1.72	6.10	7.05
Arginine	5.1	1.84	14.9	6.28
Aspartic acid	9.2	2.66	6.50	6.09
Cysteine	2.2	-	-	-
Glycine	7	1.38	5.20	6.34
Glutamic acid	10	2.76	6.90	18.24
Proline	5.2	1.08	4.90	6.92
Hydroxyproline	-	-	-	1.89
Serine	4	1.36	3.0	5.92
Tyrosine	-	0.77	1.40	4.76

Source: Adapted from Kim (2013)

3.3 Polysaccharides from Seaweeds

The seaweed cell wall mainly consists of polysaccharides and represents approximately 50% of the dry weight of seaweeds (Stiger-Pouvreau et al., 2016). The biochemical composition of these polysaccharides varies according to the class of seaweed and is influenced by several biological, physical, and environmental factors like harvesting period, seaweed species, and extraction protocol, which have a significant impact on the functional properties of the polysaccharides, as well as important structural characteristics, such as molecular weight, nature of building units, their position, type of glycosidic bond, and geometry of the molecule. Seaweed polysaccharides can be divided into two major groups, such as cell-wall polysaccharides and storage polysaccharides. Laminarins are storage polysaccharides found only in brown seaweeds, specifically in Laminariales and Fucales families. Some of these polysaccharides contain sulfate groups called sulfated polysaccharides (SP). For examples, green seaweeds contain ulvan, brown seaweeds contain fucoidans, and red seaweeds contain carrageenans. Seaweed polysaccharides, including alginate, carrageenans, fucoidans, and ulvan, are widely applied in the biological and biomedical areas due to their biocompatibility and availability (Stiger-Pouvreau et al., 2016). However, their potential application and importance as nutraceuticals and functional food ingredients are yet to be explored in detail.

3.3.1 Types of Seaweed-Derived Polysaccharides

Alginates

Alginates are polysaccharides found in the cell wall (18%–40%, dry weight basis) of brown seaweeds, such as *Ascophyllum nodosum* and *Laminaria* sp. (Rioux & Turgeon, 2015). Alginates are composed of two hexuronic acids, such as mannuronic and guluronic acids (Sánchez-Muniz et al., 2013). Alginates are used in the food industry as a gelling agent, thickening agent, stabilizer, emulsifying agent, encapsulation, and food coating material. Moreover, the beneficial health effects of alginates have been reported on human colonic microflora, reduction of toxicity of colonic luminal contacts, intestinal absorption rate, plasma cholesterol, glycemeic and insulinaemic responses (Venugopal, 2020). Furthermore, alginate-based composite biomaterials for bone-tissue regeneration are promising (Kim, 2013).

Ulvan

Ulvan is a water-soluble sulfated polysaccharide present in green seaweeds such as *Ulva* and *Enteromorpha* species. Their cell wall contains 8%–29% (dry weight basis) of ulvan (Lahaye & Robic, 2007) and is mainly composed of l-rhamnose, d-xylose, d-glucose, and d-glucuronic acid (Rioux & Turgeon, 2015). Ulvan can be extracted with water at 80–90°C with ammonium oxalate and precipitates with ethyl alcohol (Lahaye & Robic, 2007).

Fucoidan

Fucoidans are sulfated polysaccharides mainly composed with l-fucose in brown seaweeds. Especially, cell walls of seaweed species in the families Fucaceae and Laminariaceae contain 2%–10% (dry weight basis) of fucoidans. Recent studies have demonstrated the antiproliferative, antiangiogenic, and anticancer properties of fucoidan, and its importance as a marine anticancer agent in preclinical development has been presented previously (Jang et al., 2018). Moreover, fucoidan inhibits lipid accumulation in differentiated 3T3-L1 adipocytes (Gatenby et al., 2003). In the biomedical field, fucoidan will be a promising ingredient in bone-tissue engineering (Lowe et al., 2016).

Laminarin

Laminarin is composed of (1,3)-β-D-Glucan with β-(1,6) branching (Rioux & Turgeon, 2015). In addition, there are two types of laminarin chains with mannitol and glucose. The molecular weight of laminarin is around 5000 Da. Laminarin is known to have antimicrobial and anti-inflammatory activities when supplemented maternally or directly to the piglets (Heim et al., 2015).

Agar

Agar is present in red seaweeds and it is widely extracted from *Gelidium* and *Gracilaria* species. Generally, the cell wall of these red seaweeds contains 20% (dry weight basis) of agar. In addition, *Pterocladia* and *Gelidiella* species are used to extract agar in some countries. Agar consists of Ca, Mg, K, and Na sulfated esters of d- and l-galactose units. Interestingly, agar is a key alternative to gelatin (which comes from animal origin) in the jelly and confectionary industries.

Carrageenans

Carrageenans are naturally occurring sulfated linear polysaccharides found in red seaweeds. There are three types of carrageenans, k-, ι-, and λ-carrageenans. They are commonly extracted from the red species *Kappaphycus alvarezii* and *Chondrus crispus* and their cell walls contain 30%–80% (dry weight basis) of carrageenans. Ammonium, Ca, Mg, K, and Na sulphated esters of d-galactose and (3, 6)-anhydro-d-galactose units are responsible for the polysaccharide structure of carrageenans. Biological properties, chemical modification, and structural analysis of carrageenans have been reviewed previously (Tanna & Mishra, 2019). A fermented food "tofu" prepared with k/ι-hybrid carrageenans showed the highest rheological properties and carrageenan could be a practical food additive to modify the food textures (Muthumeenal et al., 2017).

All these evidences suggest that polysaccharides derived from seaweeds have a promising potential to be used as anticoagulant, antioxidant, anticancer, and antihyperlipidemic agents in the food industry as novel nutraceutical and functional food products.

3.4 Seaweed-Based Vitamins

Vitamins are organic essential compounds needed in the human body in trace amounts for different chemical and physiological processes. Vitamins are commonly classified into two groups according to their solubility: water-soluble vitamins (members of the vitamin B group and vitamin C) and fat-soluble vitamins. Even though vitamins are required only in very small quantities, to ensure that the adequate intake of vitamins in the diet is received, people can consume foods enriched with vitamins, for example, in the form of functional foods with vitamins as nutraceuticals. In addition, certain vitamins extracted from natural sources such as seaweeds have antioxidant activity and other health benefits such as decrease of blood pressure, prevention of cardiovascular diseases, or reduction of the risk of cancer (Škrovánková, 2011).

3.4.1 Importance of Vitamins for Humans

Vitamins are essential substances which cannot be synthesized by humans or only in limited quantities; therefore, they should be obtained from human diet. Vitamin deficiency can be caused not only by insufficient intake from foodstuffs but also because of increased requirements by certain groups of people (people on a special diet, smokers), poor absorption, or inadequate utilization. Seaweeds are generally a good source of some B group vitamins (B1, B2, B12). Other vitamins of B-complex are present too, but only in low or trace amounts (niacin, B6, biotin, folates). Certain seaweeds contain great quantities of vitamins with antioxidant activity, vitamins C and E, and the pro-vitamin forms of vitamin A, carotenoids.

3.4.2 Function of Vitamins in the Human Body

The hydro-soluble vitamins are needed as enzyme cofactors. The vitamins can have one or a few very specific roles or much more extensive roles. Several B group vitamins serve as coenzymes for enzymes with functions in the catabolism of foodstuffs to produce energy for the body. Some of them are fundamental even for their antioxidant activity and other health benefits. Thus low levels of some B group vitamins (B2, B6, B12) can result in reduced levels of DNA methylation and therefore in some kinds of cancer (Škrovánková, 2011).

The following vitamins are presented in seaweeds in great amounts—thiamin, riboflavin, cobalamin, and ascorbic acid. Thiamin (vitamin B1) has a key role in the intermediary carbon metabolism and is essential for several enzymes such as pyruvate dehydrogenase, pyruvate decarboxylase, and transketolase. Riboflavin (vitamin B2) is used ubiquitously throughout the cell. Cobalamin (vitamin B12) is required for the activity of cobalamin-dependent biosynthetic enzymes: methionine synthase and Methylmalonyl-CoA mutase. Interestingly, although vitamin B12 is not found in vascular plants, it is abundant in algae. As only prokaryotes have the ability to synthesize cobalamin, all of the vitamin B12 found in algae must have originally been produced by bacteria (Croft et al., 2006). Ascorbic acid (vitamin C) is a fundamental antioxidant in the ascorbate glutathione pathway. Moreover, it protects enzymes that have prosthetic transition metal ions and is a cofactor of enzymes such as violaxanthin de-epoxidase, ascorbate oxidase, and ascorbate peroxidase. Fat-soluble Vitamin E (tocopherols and tocotrienols) is an important liposoluble antioxidant which is conclusive for the prevention of oxidation of polyunsaturated fatty acids absorbed from the diet. Tocopherols block the production of reactive oxygen species formed during oxidation and help inhibit the low-density lipoprotein (LDL) oxidation as, namely, oxidatively modified LDL are considered to play a vital role in the development of atherosclerosis. Vitamin A, retinal, as visual pigments' chromophore, is important in the vision process. Besides the epithelial tissue maintenance and prevention of its keratinization, vitamin A also presents important systematic functions in growth and reproductive efficiency (Dias et al., 2011). However, plant food such as algae does not contain intrinsic vitamin A, but its provitamins, carotenoids, linear polyenes with a cyclic structure which possess a b-ring. The most abundant carotenoid with provitamin function in seaweed is β-carotene. It can be cleaved by a β-Carotene-15, 15'- dioxygenase, resulting in the formation of retinal (DellaPenna & Pogson, 2006; Dias et al., 2011).

3.4.3 Vitamin Composition of Seaweed

There are only few published data concerning vitamin content and bioavailability of vitamin forms contained in edible seaweeds. Generally, seaweeds contain both water- and fat-soluble vitamins, B-complex vitamins, vitamin C, provitamin A, and vitamin E, but some of them only in relatively low content.

The content of some vitamins in seaweeds, for example, vitamin B12 (Yamada et al., 1996), varies greatly among samples of the same species. In many cases, light is an important regulator of vitamin biosynthesis; thus plants growing in bright light have higher ascorbate content (Smith et al., 2007). Moreover, algae growing in the littoral zone or on the surface tend to have higher levels of vitamin C than algae which is harvested from depths from 9 to 18 m (Norris et al., 1937). Further, other environmental parameters, such as concentration of certain compounds in the sea, can play an important role for vitamin occurrence in algae. The importance of vitamins to plants themselves is often overlooked, but they play essential roles in plant metabolism too (Smith et al., 2007). Some algal species require different combinations of certain vitamins such as vitamins B12 and B1. Because the concentration of these vitamins in the natural environment is quite low, their absorption is insufficient (Croft et al., 2006). According to Yamada et al. (1996), red algae *Porphyra tenera* can take up the free (not protein-bounded) form of vitamin B12 from the incubation medium by concentration- and temperature-dependent processes. The amount of uptake increases with the time of incubation.

Loss of vitamins can be induced by storage conditions such as the influence of light and oxygen. Moreover, there is a negative influence on vitamin content caused by technological processing such as drying (sun-, oven-, freeze-drying) and sterilization, and culinary processes such as cooking, roasting, or baking, which could decline vitamin content due to water extrusion and high temperature during these procedures. This was observed, for example, in instable ascorbic acid (Norris et al., 1937). There are significant differences of certain vitamins' content caused by seasonal variations, for example, in *Eisenia arborea*. It was observed that the highest content of some vitamins (A, B1, B2, and partly also vitamin C) was in the spring as opposed to vitamin E, which was lowest in spring. Moreover, seasons were observed to affect carotenes' content also in *Palmaria*. The highest content of carotenes was found in the summer and was the lowest in winter (Holdt, n.d., 2010).

It is said that 100 g of seaweed provides more than the daily requirements of vitamin A, B2, B12, and two-thirds of the vitamin C requirement (Škrovánková, 2011). Most of the red seaweeds (*Palmaria, Porphyra*) contain large amounts of provitamin A and significant quantities of vitamins B1, B2, and B12, which are also present in green seaweeds. The vitamin content of brown seaweeds (*Undaria, Laminaria*) appears to be less remarkable, but brown seaweeds have high content of vitamin C (Mabeau & Fleurence, 1993). Some seaweed (such as *Porphyra*) can supply an adequate amount of vitamin B12 in vegans. Vitamin B1 and B2 are present in sufficient amount especially in brown and red marine algae. The highest amount of both vitamins was detected in wakame and kombu—0.3 and 0.24 mg B1/100 g Dw; 1.35 and 0.85 mg B2/100 g DW, respectively. Lower levels of these vitamins are present in arame (0.06–0.12 and 0.65–0.92 mg/100 g dw, respectively), *Caulerpa lentillifera* and *Ulva reticulata* (Škrovánková, 2011). The intake of vitamin B12 in strict vegetarian and vegan diets is usually quite low. Therefore, individuals can easily become deficient in this vitamin. Lower levels of vitamin B12 in a diet may result in reduced levels of DNA methylation or elevated levels of homocysteine, which is a risk factor for cardiovascular diseases (Hernandez et al., 2003). A particularly rich dietary source of the vitamin for vegans is seaweed, foods enriched with them, or seaweed extracts. Thus, consumption of some seaweed (Ex. Nori) may keep vegans from suffering B12 deficiency. The highest content of vitamin B12 in seaweed is presented in red *Porphyra* sp. (Nori)—133.8 mg B12/100 g Dw, in the form active for humans (Miyamoto et al., 2009). Different studies showed variation in B12 concentration, ranging from 12.02–68.8 mg/100 g Dw. High content is found in green laver *Enteromorpha* sp., followed by dulse, and low levels in *Ulva* sp., Wakame, Kombu, and Hijiki (Škrovánková, 2011).

Vitamin C is present especially in brown and green seaweeds, less in red algae. The highest levels of vitamin C were discovered in *Enteromorpha flexuosa* and *Ulva fasciata* (300 and 220 mg/100 g dw, respectively) (McDermid & Stuercke, 2003). The FIP states that the highest level of vitamin C is found in wakame (184.75 mg/100 g dw), red laver, and sea lettuce (MacArtain et al., 2007). Chan et al. (1997) determined the high content of vitamin C in freeze-dried algae *Sargassum hemiphylum*—153.8 mg/100 g dw—and a smaller amount in oven- and sun-dried seaweed. Other seaweeds contain much less vitamin C; high content is found in red algae *Kappaphycus alvarezzi* (107.1 mg/100 g dw) and low levels in arame, ogonori, *Sargassum polycystum, Eucheuma cottonii, U. reticulata*, and *C. lentillifera* (Fayaz et al., 2005). Despite the low lipid content in seaweed, their fat contains high levels of vitamin E. Generally, brown seaweeds contain more α-tocopherol (also β- and γ-tocopherols) than red and green algae which contain only α-tocopherol. The highest amount of vitamin E was detected in kelp *Macrocystis pyrifera*, 132.77 mg/100 g fat (α-tocopherol), with total content of 145.72 mg/100 g fat, and in *Ulva lactuca* with γ-tocopherol value 96.35 mg/100 g fat, so compared to traditional plant oils, the fat of these seaweeds contains high levels of tocols. Plant food such as algae does not contain intrinsic vitamin A, but its provitamins such as β-carotene. The high values of β-carotene with vitamin A activity were found in red seaweeds *Gracilaria changgi, K. alvarezzi,* and in brown algae kombu (5.2, 5.26, and 2.99 mg/100 g dw, respectively, recalculated to 865, 865, and 481 RE/100 g dw). Moderate levels of vitamin A were determined in wakame, arame, sea grapes, and sea lettuce (Škrovánková, 2011).

3.5 Minerals from Seaweed

Minerals are significantly important elements that perform many necessary functions in the living body, including cell transport and a wide range of metabolic processes serving as various catalytic metalloenzymes cofactors.

Edible seaweeds have high nutritional value as a source of polyunsaturated fatty acids, proteins, carbohydrates, vitamins, and minerals. Furthermore, they have been widely utilized in nutritional supplements and health foods. Human consumption of seaweeds varies by nation (Lozano Muñoz & Díaz, 2020). In Japan, the annual per capita consumption was 9.6 g of seaweed per day as of 2014. The global use of seaweeds for human consumption is increasing due to their contribution to health and their wider use as food additives (Wells et al., 2016). Seaweeds contain 10 to 20 times the amount of minerals when compared to land plants. The minerals are gathered from seawater, which makes seaweed rich in macro-elements and trace elements (Mišurcová et al., 2011). The human body requires a certain amount of some minerals to function properly, and can affect different aspects of human health. Minerals like Ca, Cu, Fe, Mg, Zn, K, Na, P, Se, Mn, and Cr are considered to be essential for human health (Campbell, 2001; Lozano Muñoz & Díaz, 2020).

3.5.1 Mineral Composition of Green Seaweeds

The mineral composition of green seaweeds has been reviewed for 23 edible Chlorophyta species: *Caulerpa lentillifera, Caulerpa racemosa, Caulerpa sertularioides, Caulerpa taxifolia, Cladophora albida, Cladophoropsis vaucheriaeformis, Codium bursa, Codium fragile, Codium indicum* (formerly *Codium iyangarii*), *Codium reediae, Codium yezoense, Dasycladus vermicularis, Ulva flexuosa* (formerly *Enteromorpha flexuosa*), *Ulva intestinalis* (formerly *Enteromorpha intestinalis*), *Ulva prolifera* (formerly *Enteromorpha prolifera*), *Monostroma hariotti, Gayralia oxysperma* (formerly *Monostroma oxyspermum*), *Ulva lactuca* (formerly *Ulva fasciata*), *Ulva fenestrata, Ulva australis* (formerly *Ulva pertusa*), *Ulva reticulata, Ulva rigida,* and *Ulvaria splendens* (Lozano Muñoz & Díaz, 2020; Pereira, 2016).

The mineral composition at 1 gram dry matter is shown in Table 3.3. For all the seaweed classes, K (3.1–27 mg/g Dw), Ca (6.9–25.3 mg/g Dw), Mg, and

Table 3.3 Essential Minerals Composition of Edible Seaweed

Scientific Name	K	Ca	Na	P	Cu	Fe	Se	Mn	Zn	Mg	Cr	I
							Essential Minerals mg/g dw					
							Cu, Se, Cr, and I µg/g dw					
Brown Seaweed (Phaeophyceae)												
Laminaria digitata	115.7	10.05	38.1	3	2.9	0.047	0.026	0.002	0.017	6.5	1.3	2000
Saccharina japonica	96.3	12.7	29.2	4.8	<0.5	0.080	8.0	0.004	0.018	6.4	1.0	2100
Saccharina latissimi					<0.5	0.040					<0.5	230
Macrocystis pyrifera	118	37.9	41.2	7.8	0.92	0.267	31.7	0.007	0.700	10.6	0.7	2100
Sargassum fusiforme	52.63	18.6	32.9	1.16	1	0.886	10	0.001	0.013	13.46	0.55	430
Undaria pinnatifida	4.8	8.9	98.4	3.6	4.3	0.184	8	0.0075	0.032	8.68	55	139
Green Seaweed (Chlorophyceae)												
Caulerpa spp.	3.18	18.5	25.74	0.29	8	0.813		0.04	0.06	3.8	3.1	
Codium fragile	14.2	25.3	92.3	5		9.43				15.2	16.8	154
Ulva clathrata (formerly Enteromorpha clathrata)	27.0	8.0	4.0	30.0	7.5	1.712	0.41	0.051	0.18	35.0	1.05	7530
Monostroma nitidium	8.1	6.9	18.0	2.0		0.025						63
Red Seaweed (Rhodophyceae)												
Gracilaria spp.	34.17	4.02	54.65	18.2	8	0.036			0.043	5.65		4260
Palmaria palmata	81.0	3.8	10	5	3.7	0.717		0.011	0.037	1.6	0.98	72
Porphyra tenera	35	3.9	36.2	2	15.8	1.832		0.36	0.02	5.65	2	185
Porphyra/Pyropia spp.	27.2	3.39	5.87	5.1	9.5	0.383	0.16	0.012	0.383	3.5	1.64	35.8

Source: Adopted from Lozano Muñoz and Diaz (2020)

Fe are present in all green seaweed. The highest K value occurred in *Ulva clathrata*, and the highest Ca value occurred for *C. fragile*. Na had a wide range of concentrations (4.0–92.3 mg/g Dw), with the highest value in *C. fragile* and the lowest value in *Ulva clathrata*. P also had a wide range (0.29–30 mg/g Dw), with no P found for *C. fragile* and the highest value found for *Ulva clathrata*. Data for Cu were found for only two seaweeds: *Caulerpa* spp. (8 mg/g Dw) and *Ulva clathrata* (7.5 mg/g Dw). The range of Fe in green seaweeds was 0.025–9.43 mg/g Dw, with the highest value in *C. fragile*. Data for Se were only found for *Ulva clathrata* with a value of 0.41 mg/g Dw. Mn is present in *Caulerpa* spp. (0.04 mg/g Dw) and *Ulva clathrata* (0.051 mg/g Dw). Zn values were found for only *Caulerpa* spp. (0.06 mg/g Dw) and *Ulva clathrata* (0.18 mg/g Dw). The range of Mg content was 3.8–35 mg/g Dw, with the highest value in *Ulva clathrata* and no values found for *M. nitidium*. The Cr content was 1.0–16.9 mg/g Dw, with the highest value in *C. fragile*. Green seaweeds had low values for I (0.063–0.154 mg/g Dw) with the exception of *Ulva clathrata* (7.53 mg/g Dw). Compared with brown and red seaweeds, the selected types of edible Chlorophyta have low content of the essential minerals K (3.18–14.2 mg/g Dw), Zn (0.06–18 mg/g Dw), and Cu (7.5–18 mg/g Dw) with the exception of *Ulva clathrata*, which had a high content of K (27.0 mg/g Dw). Green seaweeds have high content of Ca and Na with the exception of *Ulva clathrata*, which has a low content of Na (4.0 mg/g Dw). The Na content was highest in *C. fragile*. *Ulva clathrata* has a high content of I (7530 mg/g Dw) and Mg (35 mg/g Dw), and *C. fragile* has a high content of Fe of 9.43 mg/g Dw. Among green seaweeds, *Ulva clathrata* had the most complete profile of minerals but had low levels of sodium content compared to the potassium content.

3.5.2 Mineral Composition of Rhodophyceae

The content of the essential minerals K, Na, Ca, Mg, and P in edible red seaweed has been reported for 29 species or genera: *Chondrus crispus*, *Gracilaria* spp., *Palmaria palmata*, *Pyropia tenera* (formerly *Porphyra tenera*), *P. umblicalis*, *P. yezoensis*, *Gymnogongrus durvillei* (formerly *Ahnfeltiopsis concinna*), *Ceramium boydenii*, *Ceramium kondoi*, *Chondrus ocellatus*, *Corallina pilulifera*, *Eucheuma denticulatum*, *Gelidium amansii*, *Gloiosiphonia capillaris*, *Gracilariopsis longissimia* (formerly *Gracilaria confervoides*), *G. coronopifolia*, *G. parvispora*, *G. salicornia*, *Halymenia formosa*, *Hyalosiphonia caespitosa*, *Kappaphycus alvarezii*, *Laurencia okamurae*, *Leathesia marina* (formerly *Leathesia difformes*), *Myelophycus simplex*, *Palmaria* spp., *Polysiphonia stricta* (formerly *Polysiphonia urceolate*), *Porphyra/Pyropia* spp., *P. vietnamensis*, and *Rohodomela confervoides* (Moreda-Piñeiro et al., 2011; Pereira, 2016). The contents of ultra-trace elements Al, As, Au, Ba, Br, Cd, Co, Cr, Cu, Fe, Hg, and I have been reported for 25 edible red seaweeds species and the two genera *Palmaria* and *Porphyra/Pyropia* (Lozano Muñoz & Díaz, 2020).

3.5.3 Mineral Composition of Brown Seaweed

Brown seaweeds contain reserves of complex polysaccharides and higher alcohols, and their natural habitat is mostly the coastal areas in cold water bodies. Brown seaweeds have higher absorption rates than green and red seaweeds due to the presence of alginic acid, alginate, and alginic acid salt. These polysaccharides have an affinity with calcium, magnesium, sodium, and potassium salts. Brown seaweeds are also a significant source of iodine, especially the genus *Laminaria*, which has a great capacity to accumulate iodine at more than 30,000 times the iodine concentration in seawater (Lozano Muñoz & Díaz, 2020).

The composition of the essential minerals K, Na, Ca, Mg, and P in edible brown seaweeds has been reported for 24 species: *Fucus vesiculosus, Himantalia elongata, Laminaria digitata, Saccharina japonica* (formerly *Laminaria japonica*), *S. latissima, Sargassum fusiforme* (formerly *Hizikia fusiformis*), *Undaria pinnatifida, Colpomenia sinuosa, Desmarestia viridis, Dictyota acutiloba, Dictyota sandivicensis, Laminaria ochroleuca, Padina australis, Punctaria plantaginea, Sargassum carpophyllum, Sargassum aquifolium* (formerly *Sargassum echinocarpum*), *Sargassum henslowianum, Sargassum miyabei* (formerly *Sargassum kjellmanianum*) *Sargassum obtusifolium, Sargassum parvifolium, Sargassum polycystum, Sargassum thunbergii, Sargassum vachellianum*, and *Turbinaria* (Lozano Muñoz & Díaz, 2020; Pereira, 2016).

For Na, the brown seaweeds had similar values (29.2–41.2 mg/g dw) with the exception of *Undaria pinnatifida*, which had a high Na value (98.4 mg/g dw). The phosphorus content was also similar (1.16–4.8 mg/g dw) with the exception of *M. pyrifera*, which had a higher value (7.8 mg/g dw). Fe had similar values for the genus *Laminaria* (0.040–0.080 mg/g dw) and higher value for *Sargasum fusiforme* (0.886 mg/g dw), followed by *M. pyrifera* (0.267 mg/g dw) and *U. pinnatifida* (0.184 mg/g dw). There was high variability in the content of Cu in brown seaweeds (<0.5–4.3 mg/g dw), with a higher value in *U. pinnatifida* (4.3 mg/g dw), followed by *L. digitata* (2.9 mg/g dw), and with low presence in *Saccharina japonica* and *Saccharina latissima*. Se was high in brown seaweeds compared to the other groups with the exception of *Laminaria digitata*. The highest value for Se was in *M. pyrifera* (31.7 mg/g dw). The brown seaweeds presented a wide range of Cr content (<0.5–55 mg/g dw) with a very high value for *U. pinnatifida* (55 mg/g dw) for this mineral. Iodine had a high presence in this group, with the lowest values occurring for *U. pinnatifida* (13,955 mg/g dw) and highest values occurring for the genus *Laminaria* and *M. pyrifera*, where it reached up to 2100 mg/g dw. For magnesium and manganese, there were similar values in the whole group. Brown seaweeds have low manganese content and high magnesium content in comparison to red and green seaweed. The Mg content was highest in *S. japonica* (Lozano Muñoz & Díaz, 2020; Pereira, 2016).

References

Areche, C., San-Martín, A., Rovirosa, J., Soto-Delgado, J., & Contreras, R. (2009). An unusual halogenated meroditerpenoid from Stypopodium flabelliforme: Studies by NMR spectroscopic and computational methods. *Phytochemistry*, *70*(10), 1315–1320. https://doi.org/10.1016/J.PHYTOCHEM.2009.07.017

Campbell, J. D. (2001). Lifestyle, minerals and health. *Medical Hypotheses*, *57*(5), 521–531. https://doi.org/10.1054/MEHY.2001.1351

Černá, M. (2011). Seaweed proteins and amino acids as nutraceuticals. In *Advances in Food and Nutrition Research* (1st ed., Vol. 64). Elsevier Inc. https://doi.org/10.1016/B978-0-12-387669-0.00024-7

Chan, J. C. C., Cheung, P. C. K., & Ang, P. O. (1997). Comparative studies on the effect of three drying methods on the nutritional composition of seaweed sargassum hemiphyllum (Turn.) C. Ag. *Journal of Agricultural and Food Chemistry*, *45*(8), 3056–3059. https://doi.org/10.1021/jf9701749

Chatter, R., Kladi, M., Tarhouni, S., Maatoug, R., Kharrat, R., Vagias, C., & Roussis, V. (2009). Neorogioltriol: A brominated diterpene with analgesic activity from Laurencia glandulifera. *Phytochemistry Letters*, *2*(1), 25–28. https://doi.org/10.1016/j.phytol.2008.10.008

Croft, M. T., Warren, M. J., & Smith, A. G. (2006). Algae need their vitamins. *Eukaryotic Cell*, *5*(8), 1175–1183. https://doi.org/10.1128/EC.00097-06

DellaPenna, D., & Pogson, B. J. (2006). Vitamin synthesis in plants: Tocopherols and carotenoids. *Annual Review of Plant Biology*, *57*, 711–738. https://doi.org/10.1146/annurev.arplant.56.032604.144301

Denis, C., Morançais, M., Li, M., Deniaud, E., Gaudin, P., Wielgosz-Collin, G., Barnathan, G., Jaouen, P., & Fleurence, J. (2010). Study of the chemical composition of edible red macroalgae Grateloupia turuturu from Brittany (France). *Food Chemistry*, *119*(3), 913–917. https://doi.org/10.1016/j.foodchem.2009.07.047

Dere, Ş., Dalkiran, N., Karacaoğlu, D., Yildiz, G., & Dere, E. (2003). The determination of total protein, total soluble carbohydrate and pigment contents of some macroalgae collected from Gemlik-Karacaali (Bursa) and Erdek-Ormanli (Balikesir) in the Sea of Marmara, Turkey. *Oceanologia*, *45*(3), 453–471.

Dias, B., Daniel, R., Barreto, W., Alice, M., & Coelho, Z. (2011). Technological aspects of β-carotene production. *Springer*, *4*(5), 693–701. https://doi.org/10.1007/s11947-011-0545-3

Fayaz, M., Namitha, K. K., Murthy, K. N. C., Swamy, M. M., Sarada, R., Khanam, S., Subbarao, P. V., & Ravishankar, G. A. (2005). Chemical composition, iron bioavailability, and antioxidant activity of Kappaphycus alvarezzi (Doty). *Journal of Agricultural and Food Chemistry*, *53*(3), 792–797. https://doi.org/10.1021/JF0493627

Gatenby, C. M., Orcutt, D. M., Kreeger, D. A., Parker, B. C., Jones, V. A., & Neves, R. J. (2003). Biochemical composition of three algal species proposed as food for captive freshwater mussels. *Journal of Applied Phycology*, *15*(1), 1–11. https://doi.org/10.1023/A:1022929423011

Gressler, V., Fujii, M. T., Martins, A. P., Colepicolo, P., Mancini-Filho, J., & Pinto, E. (2011). Biochemical composition of two red seaweed species grown on the Brazilian coast. *Journal of the Science of Food and Agriculture*, *91*(9), 1687–1692. https://doi.org/10.1002/jsfa.4370

Gressler, V., Yokoya, N. S., Fujii, M. T., Colepicolo, P., Filho, J. M., Torres, R. P., & Pinto, E. (2010). Lipid, fatty acid, protein, amino acid and ash contents in four Brazilian red algae species. *Food Chemistry*, *120*(2), 585–590. https://doi.org/10.1016/j.foodchem.2009.10.028

Heim, G., Sweeney, T., O'Shea, C. J., Doyle, D. N., & O'Doherty, J. V. (2015). Effect of maternal dietary supplementation of laminarin and fucoidan, independently or in combination, on pig growth performance and aspects of intestinal health. *Animal Feed Science and Technology*, *204*, 28–41. https://doi.org/10.1016/j.anifeedsci.2015.02.007

Hernandez, B. Y., McDuffie, K., Wilkens, L. R., Kamemoto, L., & Goodman, M. T. (2003). Diet and premalignant lesions of the cervix: Evidence of a protective role for folate, riboflavin, thiamin, and vitamin B12. *Cancer Causes and Control, 14*(9), 859–870. https://doi.org/10.1023/B:CACO.0000003841.54413.98

Holdt, L. S., & Kraan, S. (2010). Bioactive compounds in seaweed: Functional food applications and legislation. *Journal of Applied Phycology, 23*, 543–597.

Jang, E. J., Kim, S. C., Lee, J. H., Lee, J. R., Kim, I. K., Baek, S. Y., & Kim, Y. W. (2018). Fucoxanthin, the constituent of Laminaria japonica, triggers AMPK-mediated cytoprotection and autophagy in hepatocytes under oxidative stress. *BMC Complementary and Alternative Medicine, 18*(1), 97. https://doi.org/10.1186/s12906-018-2164-2

Kim, S. (2013). *Marine Nutraceutical.* CRC Press.

Lahaye, M., & Robic, A. (2007). Structure and functional properties of ulvan, a polysaccharide from green seaweeds. *Biomacromolecules, 8*(6), 1765–1774. https://doi.org/10.1021/BM061185Q

Lee, Y. H., Yoon, S. J., Kim, A., Seo, H., & Ko, S. (2016). Health performance and challenges in Korea: A review of the global burden of disease study 2013. *Journal of Korean Medical Science, 31*(31), S114–S120. https://doi.org/10.3346/jkms.2016.31.S2.S114

Li, K., Li, X. M., Ji, N. Y., & Wang, B. G. (2007). Natural bromophenols from the marine red alga Polysiphonia urceolata (Rhodomelaceae): Structural elucidation and DPPH radical-scavenging activity. *Bioorganic & Medicinal Chemistry, 15*(21), 6627–6631. https://doi.org/10.1016/J.BMC.2007.08.023

Lowe, B., Venkatesan, J., Anil, S., Shim, M. S., & Kim, S. K. (2016). Preparation and characterization of chitosan-natural nano hydroxyapatite-fucoidan nanocomposites for bone tissue engineering. *International Journal of Biological Macromolecules, 93*, 1479–1487. https://doi.org/10.1016/J.IJBIOMAC.2016.02.054

Lozano Muñoz, I., & Díaz, N. F. (2020). Minerals in edible seaweed: Health benefits and food safety issues. *Critical Reviews in Food Science and Nutrition*, 1–16. https://doi.org/10.1080/10408398.2020.1844637

Mabeau, S., & Fleurence, J. (1993). Seaweed in food products: Biochemical and nutritional aspects. *Trends in Food Science & Technology, 4*(4), 103–107. https://doi.org/10.1016/0924-2244(93)90091-N

MacArtain, P., Gill, C. I. R., Brooks, M., Campbell, R., & Rowland, I. R. (2007). Nutritional value of edible seaweeds. *Nutrition Reviews, 65*(12), 535–543. https://doi.org/10.1301/nr.2007.dec.535-543

Matanjun, P., Mohamed, S., Mustapha, N. M., & Muhammad, K. (2009). Nutrient content of tropical edible seaweeds, Eucheuma cottonii, Caulerpa lentillifera and Sargassum polycystum. *Journal of Applied Phycology, 21*(1), 75–80. https://doi.org/10.1007/s10811-008-9326-4

McDermid, K. J., & Stuercke, B. (2003). Nutritional composition of edible Hawaiian seaweeds. *Journal of Applied Phycology, 15*(6), 513–524. https://doi.org/10.1023/B:JAPH.0000004345.31686.7F

Mišurcová, L., Machů, L., & Orsavová, J. (2011). Seaweed minerals as nutraceuticals. *Advances in Food and Nutrition Research, 64*, 371–390. https://doi.org/10.1016/B978-0-12-387669-0.00029-6

Miyamoto, E., Yabuta, Y., Kwak, C. S., Enomoto, T., & Watanabe, F. (2009). Characterization of vitamin B12 compounds from Korean purple laver (Porphyra sp.) products. *Journal of Agricultural and Food Chemistry, 57*(7), 2793–2796. https://doi.org/10.1021/JF803755S

Moreda-Piñeiro, A., Peña-Vázquez, E., & Bermejo-Barrera, P. (2011). Significance of the presence of trace and ultratrace elements in seaweeds. In S. Kim (Ed.), *Handbook of Marine Macroalgae: Biotechnology and Applied Phycology* (pp. 116–170). John Wiley & Sons. https://doi.org/10.1002/9781119977087.ch6

Murata, M., & Nakazoe, J. I. (2001). Production and use of marine algae in Japan. In *Japan Agricultural Research Quarterly* (Vol. 35, Issue 4, pp. 281–290). Springer. https://doi.org/10.6090/jarq.35.281

Muthumeenal, A., Sundar Pethaiah, S., & Nagendran, A. (2017). Biopolymer composites in fuel cells. *Biopolymer Composites in Electronics*, 185–217. https://doi.org/10.1016/B978-0-12-809261-3.00006-1

Nanri, A., Mizoue, T., Shimazu, T., Ishihara, J., Takachi, R., Noda, M., Iso, H., Sasazuki, S., Sawada, N., & Tsugane, S. (2017). Dietary patterns and all-cause, cancer, and cardiovascular disease mortality in Japanese men and women: The Japan public health center-based prospective study. *PLoS ONE*, *12*(4). https://doi.org/10.1371/journal.pone.0174848

Norris, E. R., Simeon, M. K., & Williams, H. B. (1937). The vitamin B and vitamin C content of marine algae. *The Journal of Nutrition*, *13*(4), 425–433. https://doi.org/10.1093/jn/13.4.425

Ortiz, J., Romero, N., Robert, P., Araya, J., Lopez-Hernández, J., Bozzo, C., Navarrete, E., Osorio, A., & Rios, A. (2006). Dietary fiber, amino acid, fatty acid and tocopherol contents of the edible seaweeds Ulva lactuca and Durvillaea antarctica. *Food Chemistry*, *99*(1), 98–104. https://doi.org/10.1016/j.foodchem.2005.07.027

Pangestuti, R., & Kim, S. (2015). Seaweed proteins, peptides, and amino acids. In *Seaweed Sustainability*. Elsevier Inc. https://doi.org/10.1016/B978-0-12-418697-2/00006-4

Pereira, L. (2011). A review of the nutrient composition of selected edible seaweeds. In Vitor H. Pomin (Ed.), *Seaweed: Ecology, Nutrient Composition and Medicinal Uses* (pp. 15–47). Nova Science Publishers.

Pereira, L. (2016). Edible seaweeds of the world. In *Edible Seaweeds of the World*. CRC Press. https://doi.org/10.1201/b19970

Ramos, M. V., Monteiro, A. C. O., Moreira, R. A., & Carvalho, A. D. F. A. F. U. (2000). Amino acid composition of some Brazilian seaweed species. *Journal of Food Biochemistry*, *24*(1), 33–39. https://doi.org/10.1111/j.1745-4514.2000.tb00041.x

Renaud, S. M., & Luong-Van, J. T. (2006). Seasonal variation in the chemical composition of tropical Australian marine macroalgae. *Journal of Applied Phycology*, *18*(3–5), 381–387. https://doi.org/10.1007/s10811-006-9034-x

Rioux, L. E., & Turgeon, S. L. (2015). Seaweed carbohydrates. In *Seaweed Sustainability: Food and Non-Food Applications*. Elsevier Inc. https://doi.org/10.1016/B978-0-12-418697-2.00007-6

Sánchez-Muniz, F. J., Bocanegra de Juana, A., Bastida, S., & Benedí, J. (2013). Algae and cardiovascular health. In Herminia Dominguez (Ed.), *Functional Ingredients from Algae for Foods and Nutraceuticals*. Woodhead Publishing Series in Food Science, Technology and Nutrition. https://doi.org/10.1533/9780857098689.2.369

Sen, A. K., Das, A. K., Banerji, N., Siddhanta, A. K., Mody, K. H., Ramavat, B. K., Chauhan, V. D., Vedasiromoni, J. R., & Ganguly, D. K. (1994). A new sulfated polysaccharide with potent blood anti-coagulant activity from the red seaweed Grateloupia indica. *International Journal of Biological Macromolecules*, *16*(5), 279–280. https://doi.org/10.1016/0141-8130(94)90034-5

Shannon, E., & Abu-Ghannam, N. (2019). Seaweeds as nutraceuticals for health and nutrition. *Phycologia*, *58*(5), 563–577. https://doi.org/10.1080/00318884.2019.1640533

Škrovánková, S. (2011). Seaweed vitamins as nutraceuticals. *Advances in Food and Nutrition Research*, *64*, 357–369. https://doi.org/10.1016/B978-0-12-387669-0.00028-4

Smith, A. G., Croft, M. T., Moulin, M., & Webb, M. E. (2007). Plants need their vitamins too. *Current Opinion in Plant Biology*, *10*(3), 266–275. https://doi.org/10.1016/j.pbi.2007.04.009

Stiger-Pouvreau, V., Bourgougnon, N., & Deslandes, E. (2016). Carbohydrates from Seaweeds. In Joël Fleurence and Ira Levine (Eds.), *Seaweed in Health and Disease Prevention* (Issue January 2018). Elsevier Inc. https://doi.org/10.1016/B978-0-12-802772-1.00008-7

Tanna, B., & Mishra, A. (2019). Nutraceutical potential of seaweed polysaccharides: Structure, bioactivity, safety, and toxicity. *Comprehensive Reviews in Food Science and Food Safety*, *18*(3), 817–831. https://doi.org/10.1111/1541-4337.12441

Vairappan, C. S., Anangdan, S. P., Tan, K. L., & Matsunaga, S. (2010). Role of secondary metabolites as defense chemicals against ice-ice disease bacteria in biofouler at carrageenophyte farms. *Journal of Applied Phycology, 22*(3), 305–311. https://doi.org/10.1007/s10811-009-9460-7

Venugopal, V. (2020). Seaweed: Nutritional value, bioactive properties, and uses. In Vazhiyil Venugopal (Ed.), *Marine Products for Healthcare*. CRC Press. https://doi.org/10.1201/9781420052640-13

Wells, M. L., Potin, P., Craigie, J. S., Raven, J. A., Merchant, S. S., Helliwell, K. E., Smith, A. G., Camire, M. E., & Brawley, S. H. (2016). Algae as nutritional and functional food sources: Revisiting our understanding. *Journal of Applied Phycology, 29*(2), 949–982. https://doi.org/10.1007/S10811-016-0974-5

Wong, K. H., & Cheung, P. C. K. (2000). Nutritional evaluation of some subtropical red and green seaweeds. Part I—Proximate composition, amino acid profiles and some physico-chemical properties. *Food Chemistry, 71*(4), 475–482. https://doi.org/10.1016/S0308-8146(00)00175-8

Yaich, H., Garna, H., Besbes, S., Paquot, M., Blecker, C., & Attia, H. (2011). Chemical composition and functional properties of Ulva lactuca seaweed collected in Tunisia. *Food Chemistry, 128*(4), 895–901. https://doi.org/10.1016/j.foodchem.2011.03.114

Yamada, S., Shibata, Y., Takayama, M., Narita, Y., Sugawara, K., & Fukuda, M. (1996). Content and characteristics of vitamin B12 in some seaweeds. *Journal of Nutritional Science and Vitaminology, 42*(6), 497–505. https://doi.org/10.3177/jnsv.42.497

Zerrifi, S. E. A., Khalloufi, F. El, Oudra, B., & Vasconcelos, V. (2018). Seaweed bioactive compounds against pathogens and microalgae: Potential uses on pharmacology and harmful algae bloom control. *Marine Drugs, 16*(2), 55. https://doi.org/10.3390/md16020055

4

Nutraceuticals-Based Echinodermata

Ratih Pangestuti, Lisa Fajar Indriana, Yanuariska Putra,
Idham S. Pratama, Puji Rahmadi, and Se-Kwon Kim

Contents

DOI: 10.1201/9781003128175-4

4.1 Introduction

Numerous studies correlating bioactive materials derived from marine resources and treating some chronic diseases have shown great possibilities of developing nature-based nutraceutical products (Pangestuti and Arifin 2017). The sources of natural-based nutraceuticals exist in many reservoirs and may be found in terrestrial and marine resources. Terrestrial resources include fermented, herbal, botanical and animal products; they are by far more explored than marine resources. Even though the majority of nature-based nutraceuticals products in the marketplace are of terrestrial origin, marine organisms-based products are gaining more and more attention due to their unique features and medicinal value, which are not found in terrestrial-based resources.

Among marine organisms, Echinodermata is an interesting natural source of biological active materials with numerous health benefit effects that could be used as natural ingredients in nutraceutical products. Echinodermata are the second largest phylum in deuterostomes after the chordates. This phylum contains around 7000 species which include Holothuroidea, Echinoidea, Ophuiroidea and Asteroidea. Generally, adult Echinodermata can be found in sea beds from the intertidal to abyssal zone. This phylum has a remarkable economic value. Many marine animals belonging to this phylum are known for their bioactivities as well as health benefit effects. The most popular is sea cucumber (*teripang* in Indonesian; *beche-de-mer* in French) and sea urchin gonads. Sea cucumbers are slow-moving marine organisms that live in complex environments submitted to extreme conditions; therefore, they must adapt to the new environmental conditions to survive, and therefore synthesize unique bioactive metabolites. According to the Ming dynasty report (1368–1644 BC), the sea cucumber harbored the same medicinal properties as the herb ginseng; therefore, it also called "haishen," or ginseng from the sea. Up to now, a large number of studies from different research groups all over the world have focused on isolation of bioactive molecules from Echinodermata, exploring their medicinal value and development of nutraceutical products. In Marine Bio-industry LIPI (Rep. of Indonesia), our research group is continuously developing sustainable aquaculture for Echinodermata and exploring their nutritional and medicinal value. In addition, we also develop food products and nutraceuticals from sea cucumbers (Figure 4.1).

In this contribution, first we explore the nutritional value and biological activities and possibilities of new sources of nutraceuticals. In addition, we also explore sustainable aquaculture to support the industrial potency of Echinodermata. We believe the information provided in this chapter can be useful for many research groups, academicians and students who are interested in the search for the nutritional and medicinal value of Echinodermata.

Figure 4.1 *Teripang product developed by Marine Bio-industry LIPI Rep. of Indonesia.*

4.2 Nutritional Value and Bioactivities of Echinodermata

4.2.1 Holothuroidea

Holothuroidea, sometimes also known as sea cucumbers, are a diverse group of worm-like and usually soft-bodied Echinodermata. Sea cucumbers are one of the most popular marine organisms. It shows high potential in the food, pharmaceuticals, nutraceuticals, cosmetic as well as traditional medicine industries (Pangestuti and Arifin 2017). There are approximately 1716 sea cucumber species in the world, with the greatest biodiversity being in the Asia Pacific region. Southeast Asia represents the global market hotspots for the sea cucumbers trade due to their known mega biodiversity. Many sea cucumbers are gathered for human consumption and some are cultivated in aquaculture systems. In Indonesia we can find around 350 sea cucumber species, with at least 54 species with economic value (*teripang*). Indonesia has a long history in the sea cucumber trade, with Indonesia being recorded as the oldest sea cucumber trader. Malaysia and Philippines are also important exporters of dried sea cucumber and sea cucumber-based products.

Sea cucumber is an interesting source of ingredients for nutraceutical and traditional medicine. For its nutritional value, this marine organism is characterized by low lipid content and a high amount of protein. Proximate contents of commercially important sea cucumbers are presented in Table 4.1. Several factors might affect the lipids and protein profile of sea cucumbers. These factors include type of sea cucumber species, reproductive and feeding patterns as well as environmental conditions (including temperature of their living habitat; for example, constant temperature gives a higher lipid content than fluctuating temperature) (Rasyid et al. 2020).

Sea cucumbers contain diverse bioactive compounds such as protein, bioactive peptides, collagen, triterpene glycosides, carbohydrates, phenolic compounds. Bioactive compounds from sea cucumber have been known to possess various bioactivities such as anti-cancer, antioxidant, immune-modulatory, hepatoprotective, neuroprotective, wound healing and many other properties. One of the major and most abundant compounds isolated from sea cucumbers is triterpenoid glycosides (saponins) (Bahrami, Zhang, and Franco 2014). These compounds have been well characterized for their anti-cancer activities. As an example, five compounds isolated from *Holothuria fuscocinerea*—saponin fuscocineroside A, fuscocineroside B, fuscocineroside C, pervicoside C and holothurin A—have shown cytotoxicity against human leukemia HL-60 and human hepatoma BEL-7402 cells. In addition, holothurin A3 and holothurin A4 isolated from *Holothuria scabra* were found to be cytotoxic in human

Table 4.1 Proximate Composition (%) of Sea Cucumbers

Sea Cucumber Species	Moisture	Protein	Lipid	Ash	Carbohydrates	Ref
Stichopus hermanni (dried)	10.20 ± 0.32	47.00 ± 0.36	0.80 ± 0.02	37.90 ± 0.33	-	(Wen, Hu, and Fan 2010)
Thelonata ananas (dried)	15.10 ± 0.29	55.20 ± 0.38	1.90 ± 0.01	25.10 ± 0.30	-	(Wen, Hu, and Fan 2010)
Thelonata anax (dried)	1.20 ± 0.06	40.70 ± 0.33	9.90 ± 0.27	39.20 ± 0.28	-	(Wen, Hu, and Fan 2010)
Holothuria fuscogilva (fresh)	84.34 ± 0.72	63.64 ± 4.56	1.12 ± 0.28	30.45 ± 6.79		(Fawzya et al. 2015)
Holothuria leucospilota (fresh)	81.41 ± 0.60	45.71 ± 0.20	4.60 ± 0.30	4.30 ± 0.20	44.96 ± 0.30	(Omran 2013)
Holothuria scabra (fresh)	85.76 ± 0.30	43.43 ± 0.20	5.66 ± 0.09	2.26 ± 0.15	48.65 ± 0.20	(Omran 2013)
Holothuria scabra (fresh)	84.55 ± 0.03	6.95 ± 0.04	0.78 ± 0.02	7.38 ± 0.07	0.34 ± 0.01	(Ardiansyah et al. 2020)
Holothuria atra (dried)	9.90 ± 0.01	58.20 ± 0.72	1.32 ± 0.00	31.58 ± 0.42	-	(Ibrahim et al. 2015)
Actinopyga mauritiana (dried)	11.60 ± 0.31	63.30 ± 0.43	1.40 ± 0.02	15.40 ± 0.18	-	(Wen, Hu, and Fan 2010)

epidermoid carcinoma (KB) and human hepatocellular carcinoma (Hep-G2) (Kim and Himaya 2012).

Water extract from *Stichopus hermanni* from Malaysia have been reported to promote growth and proliferation of spinal astrocytes (Patar, Jamalullail et al. 2012). In pathological cases like spinal cord injury, proliferating reactive astrocytes are proven essential for the early regeneration process, provide neuroprotective effects and preserve motor function after acute injury. Further, it was demonstrated by GC-MS results that 37% of the total *S. hermanni* water extracts were comprised of amino acids (37%) followed by hydrocarbon (21%), ester compounds (16%); the other remaining compounds consisted of phenols, alcohol groups and unidentified compounds. The 2-carbamoyl-3-methylquinoaxaline was found to be the most abundant compounds in *S. hermanni* extracts (Patar, Jaafar et al. 2012). Interestingly, quinoxaline derivatives have been reported to be involved in reducing neurological deficits and glia loss after spinal cord injury. These *S. hermanni* have the potential to be explored and developed as neuroprotective agents.

Large numbers of studies have reported the medicinal value of sea cucumber-derived bioactive compounds. However, understanding the specific structures and bioactivities relationship of bioactive compounds from sea cucumbers is still a great challenge; further there is a considerable gap in this area compared with the isolation rate of novel bioactives. Adequate clinical trials are needed in drug development of sea cucumber-derived bioactive materials. More importantly, once their biological activities and health benefit effects are demonstrated, new aspects need to be addressed such as sustainable aquaculture of sea cucumber, production of functional ingredients at industrial scale, extraction and purification of functional ingredients, quality control and scientifically demonstrated health properties.

4.2.2 Echinoidea

Echinoidea or sea urchins are a group of marine invertebrates; they belong to the phylum of Echinodermata. Morphologically, they are characterized as the mostly familiar globose forms and surrounded by spines. Sea urchins have gained attention due to their gonads' nutritional benefits and health potential. *Uni*, or sea urchin gonads, are highly valued and considered as a delicacy in cuisine. It has been reported that *uni* are considered as a good resource of proteins, lipids, polyunsaturated fatty acids, carbohydrates, carotenoids and minerals (Sun and Chiang 2015; Archana and Babu 2016; Lourenço, Valente, and Andrade 2019; Rocha et al. 2019; Shang et al. 2020). The nutritional composition of sea urchin gonads fluctuates with the seasons. In some studies, protein is the highest of the biochemical components of sea urchin gonads between 9.26% and 16% (Dincer and Cakli 2007; Mol et al. 2008; Archana and Babu 2016; Rocha et al. 2019). The nutritional composition of dry weight basis of sea urchin gonads is presented in Table 4.2.

Table 4.2 Nutritional Composition of Dry Weight Basis of Sea Urchin Gonads

Nutritional Compositions	Composition in Gonads
Proteins (%)	9.26–16
Lipids (%)	3–6
Carbohydrates (%)	1.63–3.9
Moisture (%)	70–81.94
Ash (%)	1.5–3.76
Energy (kcal/100g)	96–167

Source: Archana and Babu (2016); Rocha et al. (2019); Dincer and Cakli (2007); Mol et al. (2008); Mamelona, Saint-Louis, and Pelletier (2010)

Fatty acids composition is important for gonad quality since it affects flavor and storage characteristics. Polyunsaturated fatty acids (PUFA) were found to be the dominant fatty acids class (40 to 44.42% of fatty acids total) (Mamelona, Saint-Louis, and Pelletier 2010; Arafa et al. 2012; Svetashev and Kharlamenko 2020). Archana and Babu (2016) mentioned that PUFA contained in sea urchin gonads is significant nutritionally. PUFA such as omega-3 fatty acids and omega-6 fatty acids are important in human nutrition. It is been reported that PUFA exhibits potential biology activities such as anti-inflammatory, antioxidant, and anti-obesity properties. In addition, a diet rich in PUFA can reduce the risk of breast cancer and colorectal cancer, prevent cardiovascular diseases, maintain and improve skin health, and elevate the performance of skeletal muscle metabolism (Dydjow-Bendek and Zagoździon 2020; Schulze et al. 2020; D'Angelo, Motti, and Meccariello 2020; Balić et al. 2020; Gammone et al. 2018).

Carotenoids play an important role in egg production and development and increase the biological defense efficiency of sea urchins (Suckling, Kelly, and Symonds 2020). In several studies, echinenone has been reported as the most abundant carotenoid in sea urchin gonads, approximately more than 70% of the total pigment. The second dominant carotenoid was β-carotene. Other carotenoid pigments also were identified, such as α-carotene, lutein, amarouciaxanthin A, fucoxanthinol, zeaxanthin, canthaxanthin and astaxanthin (Rocha et al. 2019; Suckling, Kelly, and Symonds 2020; Symonds et al. 2007; Garama, Bremer, and Carne 2012). Garama et al. (2012) found that dark color gonads have a higher content of β-carotene, astaxanthin, canthaxanthin and fucoxanthinol compared to the light color gonad. Cirino et al. (2017) reported that astaxanthin extracted from the gonads of cultured sea urchin *Arbacia lixula* has an effective radical scavenging activity and possesses higher inhibition ability than α-tocopherol as much as 33% (Cirino, Brunet et al. 2017). It is known that carotenoids from marine organisms exhibit bioactive properties such as antioxidant, anti-inflammatory, antiproliferative, anti-obesity and immunomodulator properties and provitamin A activity. Hence, the carotenoids have potential uses in prevention against oxidative damages of free radicals, immune system stimulation, decrease of the risk of cancer insurgence and expansion, cardiovascular protection, and prevention of age-related diseases (Galasso, Corinaldesi, and Sansone 2017; Suckling, Kelly, and Symonds 2020).

Polysaccharides are one of the natural products that can be isolated from marine organisms. They show various pharmaceutical activities such as antioxidant, antiviral, antibacterial, anticoagulant, immunomodulator and anti-cancer effects (Ruocco et al. 2016). It was reported that polysaccharides were able to be isolated from gonads and shells of sea urchin species. The extracted polysaccharides demonstrated significant potent anti-inflammatory, analgesic activity, and gastroprotective effects. Due to its osteogenic activity, sea urchin polysaccharides may be a potential candidate for functional food application with bone health as its target (Yang et al. 2015; Jiao et al. 2015; Salem et al. 2017).

Polyhydroxynaphthoquinones, known as echinochromes or spinochromes, are secondary metabolites that are involved in the pigmentation of sea urchins. Spinochromes appear to be an important factor in sea urchins' defense mechanism, UV protection and reproduction, as well as in their antibacterial and pro-inflammatory activities. Spinochromes can be extracted from the shells and spines of sea urchins. These quinonoid pigments are known for their various biological activities—antioxidant, anti-inflammatory, antibacterial, antiviral, fungicidal, cardioprotective and neuroprotective. According to their interesting bioactive properties, they can be used as promising components for product development in the pharmacological and cosmeceutical industries (Brasseur et al. 2017; Brasseur et al. 2018; Hou et al. 2020; Nhu Hieu et al. 2020; Vasileva et al. 2021).

4.2.3 Ophiuroidea

Ophiuroidea is a class in the phylum Echinodermata that has flexible arms to slowly move on the seabed. Together with the ability to regenerate these arms, those features are the characteristics of ophiuroids or brittle stars (Czarkwiani et al. 2016). With regard to the phylum Echinodermata that is well known to have defense biochemical compounds, there are numerous saponins (triterpenoid and steroidal) that have been described in brittle stars. However, the composition in ophiuroidea is different than holothurians due to different diet, food supply and biosynthesis pathways (Drazen et al. 2008; Claereboudt et al. 2019). Saponin in ophiuroidea are modifications from lanosterol rather than parkeol that is common in holothurians (Claereboudt et al. 2019). Although ophiuroidea utilize their saponin for a defensive strategy from their predators, it has biological activity ranging from antibacterial, anti-inflammation and anti-tumor (Amini et al. 2017; Gammone et al. 2014; Wang et al. 2004).

Biological compounds of brittle stars have been reported in several studies that predominantly contain up to ten types of terpenes. Some of these terpenes, such as 10-acetoxy-18-hydroxy-2,7-dolabelladiene, dihydroxycrenulide, dictyolactone, pachydictyol and dictyol, have the ability to inhibit bacterial growth (Wang et al. 2004). In addition, laurinterol, another terpene in ophiuroidea, has been shown to suppress metabolic activity that leads to apoptosis of tumor cells. Other compounds, pacifenol and prepacifenol epoxide, have been reported to have potential benefits for pain and inflammation treatment due to their anti-inflammatory properties (Gammone et al. 2014). These two terpenes are surprisingly found in ophiuroids because these two compounds are previously reported only in sea hares and red macroalgae (Yu et al. 2017).

Previous studies described some sterols found in ophiuroidea have been reported to have biological activity (Duque et al. 1997; Claereboudt et al. 2019; Riccio, D'Auria, and Minale 1985; Gazha et al. 2016; D'Auria et al. 1995). Their polyhydroxylated sterol mono- and disulphates have immunomodulatory properties that improve neutrophils activity and antibody production. Some of them have more prominent immunomodulatory activity due to their certain chemical structure rather than their affinity level to water (Gazha et al. 2016). Moreover, sulphated steroid from brittle stars was also found to have antiviral activity (Gustafson, Oku, and Milanowski 2004; Roccatagliata et al. 1996; Roccatagliata, Maier, and Seldes 1998). In addition, quinonic pigment (specific pigment in echinoids but including ophiuroids) also possesses anti-cancer activity as well (Riccio, D'Auria, and Minale 1985; Folmer et al. 2010; Hou et al. 2018).

Other compounds, such as maculalactone and laurinterol, have been extracted also from ophiuroidea. These phenylpropanoids have remarkable biological activity in suppressing bacterial growth and even proliferation of cancer cells (Wang et al. 2004). Methanol extract of ophiuroidea has been shown to act as an anti-metastatic and anti-angiogenic and inhibit vascularization of cancer cell tissue (Baharara, Amini, and Mousavi 2015). Porphyrin also has been found in brittle stars and can be utilized in pharmaceutical industries for cancer treatment (Klimenko et al. 2021). In addition, dichloromethane extract of ophiuroidea improves the efficacy of doxorubicin during chemotherapy for a wide range of cancer types (Afzali et al. 2017; Carvalho et al. 2009). Also, polysaccharide from brittle stars is found to act as an anti-angiogenic and anti-metastatic as well, which in the future can be utilized as a natural resource for pharmaceutical industries (Baharara, Amini, and Musavi 2017; Baharara and Amini 2015). Brittle stars also have another type of curacin that is reported as curacin E, which is slightly different than curacin A that is found in cyanobacter. However, it has similar biological activity to suppress cell proliferation (Folmer et al. 2010; Ueoka et al. 2016; Swain, Padhy, and Singh 2015).

4.2.4 Asteroidea

Asteroidea are grouped into several orders such as Brisingida, Forcipulatida, Notomyotida, Paxillosida, Spinulosida, Valvatida and Velatida. More than 1500 species can be found throughout the world's oceans. Asteroidea are important members of marine benthic ecosystems; these species can be voracious predators, having significant impacts on community structure. As an example, the crown-of-thorns starfish, *Acanthaster planci* (Figure 4.2), can cause extreme detrimental effects to coral reefs, particularly during population outbreaks.

Marine organisms belonging to this class are easily characterized by their unique distinct character, a star-shaped body plan consisting of a central disc and multiple (typically five) radiating arms (Figure 4.3). Asteroidea can be distinguished from other Ophiuroidea by the structure of the arms; in asteroidea, skeletal support for the arms is provided by the ossicles of the body wall, which merge with those of the central disc.

Asteroidea are well known for their ornamental utilization. Asteroidea also have been used in Traditional Chinese Medicine to treat many diseases. In addition, many bioactive compounds with biological activity have been reported from these classes. Bioactive compounds from asteroidea are mostly steroids, steroidal glycosides and gangliosides. These components have been found to possess various bioactivities, such as anti-cancer, anti-inflammation. As an example, spine venom from *Acanthaster planci* has been reported to inhibit proliferation and induce apoptosis in human melanoma (A375.S2) cells at 1.25 µg/ml dose (Lee et al. 2014). The cytotoxicity in A375.S2 cells induces oxidative stress from the production of reactive oxygen species (ROS)

Figure 4.2 *The crown-of-thorns starfish (*Acanthaster planci*) in its natural marine habitat (Kodek Bay, Lombok West Nusa Tenggara Indonesia).*

Figure 4.3 *Blue starfish (*Linckia laevigata*) in its natural marine habitat (Kodek Bay, Lombok West Nusa Tenggara Indonesia).*

and decreases mitochondrial membrane potential. Further, the spine venom hydrolysates were purified, and active proteins (plancitoxin I protein) were isolated (Lee, Hsieh, and Hwang 2015). The spine venom induced apoptotic procedure via regulation of MAPK p 38 pathways.

Asterosaponin (astrosteriosides A, astrosteriosides B, astrosteriosides C, astrosteriosides D), psilasteroside and marthasteroside B have been isolated from the edible Vietnamese starfish *Astropecten monacanthus* (Thao et al. 2013). All steroids have been tested in vitro and showed potent inhibitory activity on inflammatory responses. These anti-inflammatory properties of asteroidea might open further applications of asteroidea in nutraceuticals as adjuvant or complementary inflammation remedies.

4.3 Commercially Available Products Based Echinodermata and Future Prospects

Currently, the nutraceutical market is mainly classified into functional foods, dietary supplements and herbs. Echinodermata have also been used for traditional medicinal purposes since ancient times. Supporting the use of Echinodermata in traditional medicine, scientific evidence shows that Echinodermata contain diverse biologically active materials that provide health benefit effects such as antioxidant, antibacterial, antifungal, anti-cancer, antiviral, anti-inflammatory, neuroprotective activities and many others. Therefore, it is not surprising that the number of Echinodermata-based nutraceuticals products are increasing every year (Figure 4.4). Echinodermata

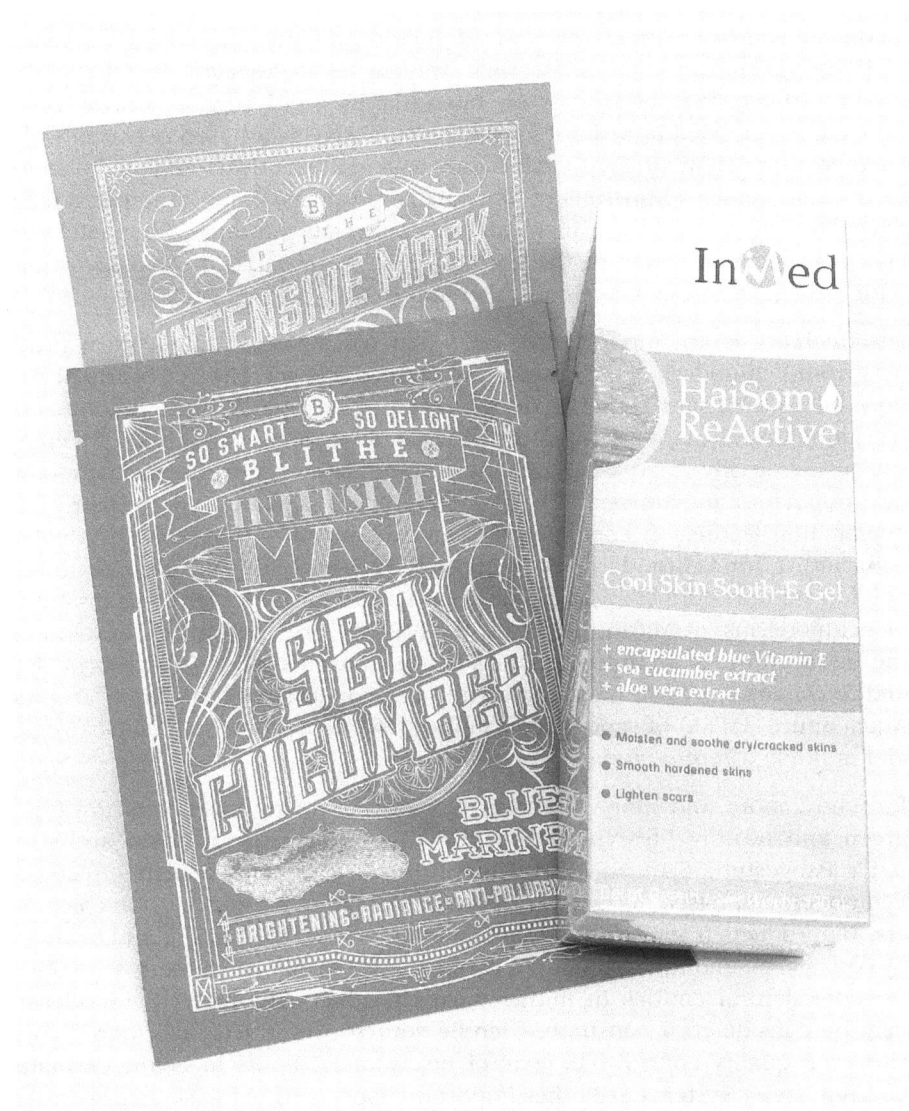

Figure 4.4 *Commercially available Echinodermata-based products in the market.*

are well recognized for their bioactive substances, which have great potential for use in nutraceutical industries. Echinodermata bioproducts possess various bioactivities and have remarkable potential to be used in the prevention and or treatment of various diseases. In addition, the great diversity of Echinodermata will open up a number of novel bioproducts extracted from Echinodermata.

Large numbers of nutraceutical products can be potentially developed from Echinodermata when we consider the number of articles and patents focusing on nutritional value, extraction, isolation of novel bioactive substances, medicinal and or health effects and product formulations. However, not many of these substances have made their way up as nutraceuticals sold in the market. Hence, further studies on bioavailability, efficacy, as well as clinical trials are needed in order to develop Echinodermata-based nutraceuticals. Further, innovation of Echinodermata-based nutraceuticals should be more focused on quality improvement of available products. As an example, a fishy flavor and odor are found in most Echinodermata products (drawbacks of Echinodermata products in the market). Hence, research on the flavor and odor reduction and/or masking is also important.

Sustainability is an important aspect when we develop nutraceutical products from Echinodermata. This marine phylum is susceptible to overexploitation due to their late maturity, density-dependent reproduction, low rates of recruitment and many of them are slow=moving organisms. Notably, many species are easily harvested, and their vulnerable nature due to their biology, population dynamics and habitat preferences all contribute to their overexploitation leading to endangerment and extinction. Therefore, sustainable aquaculture for Echinodermata is urgently required and needs to be developed to support nutraceutical industries. In order to support the sustainability of Echinoderms, at Marine Bio-industry LIPI we are continuously developing the aquaculture of Echinodermata such as sea cucumber (*Holothuria* spp and *Stichopus* spp). We have succesfully developed Integrated Multi Trophic Aquaculture (IMTA) of sand sea cucumber (*Holothuria scabra*), red seaweeds and milkfish (*Chanos chanos*).

Presently, more and more attention has been paid by customers toward a green and healthy lifestyle with natural bioproducts (Pangestuti and Kim 2017; Pangestuti, Siahaan, and Kim 2018). The health benefit effects of Echinodermata-based nutraceuticals has gained more popularity with consumers. Bioproducts from Echinodermata are considered as "natural and healthy" by consumers, and this promotes a positive response in consumers, who often consider natural entities in nutraceuticals product. Therefore, Echinoderms may be considered a consumer-friendly source of nutraceutical ingredients. However, quality control (i.e. level of heavy metal limits in Echinodermata) and regulatory systems are other important aspects to increase and maintain the market popularity for these products.

4.4 Sustainable Aquaculture of Echinodermata

Sustainable aquaculture is urgently required and needs to be developed to support development of Echinodermata in food, pharmacy and nutraceutical industries.

4.4.1 Sea Cucumber

4.4.1.1 Broodstock Management

Generally, broodstock of sea cucumber are collected from the wild by diving, snorkeling or handpicking in low-tide conditions during the reproductive season (Agudo 2006; Ren, Liu, and Pearce 2018; Domínguez-Godino and González-Wangüemert 2018; Militz et al. 2018; Laguerre et al. 2020). Selected broodstock size ranges from 250 to 1000 g in wet weight (Ren, Liu, and Pearce 2018; Militz et al. 2018) (Figure 4.5). Appropriate handling, packing and transfer method to the hatchery are crucial to minimize stress condition, evisceration and mortality rate of broodstock. Broodstock must be put in a plastic bag with oxygen-aerated sea water and then be placed in a cold insulated condition with stable temperature (Purcell, Blockmans, and Agudo 2006). On the other hand, Rakaj et al. (2019) demonstrated that dry ice could be used to

Figure 4.5 *Broodstock selection of sea cucumber* Holothuria scabra *by weight.*

maintain temperature of broodstock placed in aerated tanks. Another study by Battaglene et al. (2002) showed that sea cucumber could be transported using an insulated container without aeration; however, the water should be replaced regularly. In addition, the broodstock needs to be packed in double-layer plastic bags.

In the hatchery, broodstock are allocated to conditioning tanks in order to acclimate prior to the induction of spawning. During the rearing period, sediment is placed in the bottom of tank with recirculating aquaculture systems (RAS) (Domínguez-Godino and González-Wangüemert 2018; Rakaj et al. 2019). In addition to natural feeding from sediment, supplementary feeding such as dry powder of *Zostera noltii*, powdered *Spirulina*, fine shrimp feed is added (Duy 2012; Domínguez-Godino et al. 2015; Domínguez-Godino and González-Wangüemert 2018; Militz et al. 2018). In certain species, broodstock were maintained in the pond without adding supplementary feed (Duy 2012).

Various tropical and subtropical sea cucumber species have been cultured; nevertheless only sandfish *Holothuria scabra* has been successfully cultured extensively (Purcell, Hair, and Mills 2012). In addition, *Apostichopus japonicus* farming has risen quickly (Yuan et al. 2010). Others species are produced in hatcheries to provide juveniles, for instance *H. lesson, H. spinifera, H. fuscogilva, Parastichopus californicus, Isostichopus fuscus, Actinopyga* sp., *Australostichopus mollis, Actinopyga mauritiana, A. miliaris, Stichopus horrens* and *P. californicus* (Purcell, Hair, and Mills 2012). However, culture process and larval development are commonly similar in many holothurian species.

4.4.1.2 Spawning Induction and Fertilization

Prior to stimulation, broodstock are rinsed carefully with seawater in order to remove sediment and other organisms (Rakaj et al. 2019). Then, broodstock are placed into spawning tanks filled with UV sterilized seawater. There are various methods showing potential stimuli that have been used to trigger spawning, for instance thermal stimulation (seawater as the medium is heated to 3–5°C), high-concentrate feed stimulation, addition of live or paste microalgae, addition of dried *Schizochytrium* sp., gonad stimulation added with conspecific sperm, desiccation (Battaglene et al. 2002; Agudo 2006; Hu et al. 2013; Domínguez-Godino et al. 2015; Gianasi, Hamel, and Mercier 2019; Rakaj et al. 2019).

Broodstock show pre-spawning behavior that indicates whether the spawning process is imminent, such as climbing up the tank wall, swaying the anterior part of the body, a rolling movement (Agudo 2006; Rakaj et al. 2019), and part of the body appears to be in standing position when spawning occurs (Laguerre et al. 2020). Males eject the sperm first, then this is followed by females ejecting the mature oocytes 15–60 minutes after induction. They release gametes through gonopores located in the dorsal anterior (Rakaj et al. 2019; Laguerre et al. 2020).

Individuals are removed from the tank after the spawning activity is complete. The eggs are collected by siphoning the water and filtering it using an 80–100 μm sieve, then they are rinsed several times by UV-filtered seawater to remove excess of sperm (Domínguez-Godino and González-Wangüemert 2018; Militz et al. 2018). The quality of broodstock and success of the spawning induction influence the quality of the eggs, the abundance of oocyte release and the survival of embryos (Gianasi, Hamel, and Mercier 2019).

4.4.1.3 Embryonic Development and Larvae Rearing

A tank containing UV-filtered seawater with a mild aeration system is prepared to place the eggs collected from the broodstock, without water exchange (Militz et al. 2018). During embryonic larvae development, eggs are observed using a microscope in order to assess the fertilization success. Approximately 1 to 2 hours after fertilization, embryonic development commences with the cleavage stages (Militz et al. 2018; Laguerre et al. 2020). The fertilized eggs develop through cell division as blastula, gastrula and larvae. Sea cucumber has three defined stages of larvae development—auricularia (early, mid, late), doliolaria (early, mid, late) and pentactula.

Throughout the larval rearing process, water in the tanks is replaced daily up to 40–50% of total volume by siphoning to remove bacterial films, dead larvae and uneaten food suspended in the bottom of the tank (Domínguez-Godino et al. 2015; Laguerre et al. 2020). Water quality parameters such as temperature, salinity, pH and dissolved oxygen are monitored daily (Domínguez-Godino et al. 2015). Unicellular or mixed microalgae are provided daily to feed the larvae including *Tetraselmis chuii*, *Chaetoceros calcitrans*, *C. muelleri*, *C. salina*, *Rhodomonas salina*, *Isochrysis galbana*, *Nitzschia closterium minutissima*, *Dunaliella* sp. and *Chlorella pynenoidosa* (Mercier, Battaglene, and Hamel 2000; Li, Li, and Kong 2010; Hu et al. 2013; Ren, Liu, and Pearce 2018; Rakaj et al. 2019; Militz et al. 2018). Density of microalgae ranges from 10,000 to 15,000 cell/mL in the early auricularia stage and is raised in the following stage (Juinio-Meñez et al. 2012; Hu et al. 2013).

In the late auricularia stage, they metamorphose into non-feeding doliolaria resulting in the reduction of size, and they tend to swim near the wall (Rakaj et al. 2019). The doliolaria larvae are transferred into outdoor tanks which are filled with UV filtered seawater and are provided with the settlement plate. In the hatchery, the successful settlement of the larval phase is affected by appropriate collectors as substrate (Li, Li, and Kong 2010). The settlement process is influenced by physical, biological and chemical factors (Li, Li, and Kong 2010). A material shelter is made from PVC undulated panels, *Spirulina*-coated plates, polyethylene plastic sheet with diatom-based films and plastic sheet with natural microbial film (Li, Li, and Kong 2010; Domínguez-Godino et al.

2015; Altamirano and Noran-Baylon 2020). Mercier, Battaglene, and Hamel (2000) reported about the induction of the settlement phase of *H. scabra* using seagrass leaves *Thalassia hemprichii*. Doliolaria larvae transform into benthic pentactula (Domínguez-Godino and González-Wangüemert 2018). In post-settled juveniles, diatom *Navicula ramossissima* and *Sargassum* extract are given as supplementary food (Juinio-Meñez et al. 2012). Juvenile *H. scabra* settled successfully on mesh that is coated by natural biofilm (Figure 4.6).

The development of sea cucumber from fertilization to being a juvenile is different in each stage. *Stichopus* sp., known as curry fish, reaches its juvenile stadia of 400 µm in 12 days (Hu et al. 2010). For instance, *S. horrens* manages to reach its juvenile stadia of 1682 µm in 30 days (Hu et al. 2013). Meanwhile, *H. forskali* gains its visible juvenile stadia within two months (Laguerre et al. 2020).

4.4.1.4 Nursery and Juvenile Grow-Out

In the laboratory scale, *H. scabra* juveniles are reared in the nursery tanks containing sand-filtered seawater and sediment enriched with *Sargassum* and diatoms as their diets (Juinio-Meñez et al. 2012). The juveniles of *H. scabra* are moved into the nursery tank when they reach 5 mm length size in the

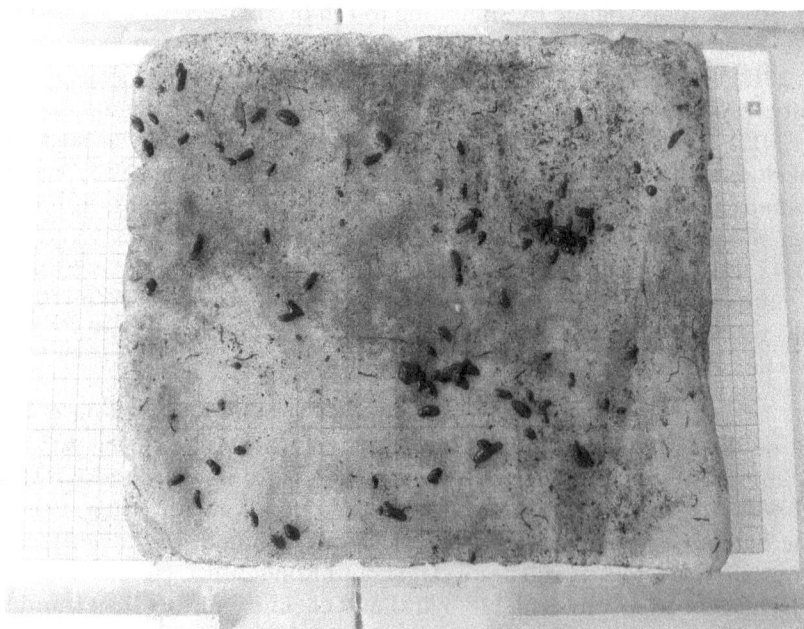

Figure 4.6 *Juveniles of* Holothuria scabra *settled on mesh that is coated by natural biofilm.*

settlement tank (Mills et al. 2012). In terms of massive scale production, it is complicated to carry out the nursery of *H. scabra* in a laboratory or hatchery area due to limited availability of food, cost and equipment (Juinio-Meñez et al. 2012; Juinio-Meñez et al. 2017). Post settlement, larvae that have reached the early juvenile stage then can be maintained in the nursery and grow out in the wild (Figure 4.7). According to Altamirano and Noran-Baylon (2020), good management practice in the nursery contributes to the increase of sea cucumber production for both restocking and aquaculture purposes.

Various nursery systems have been developed in pond- and sea-based rearing systems. The ocean nursery system has been designed to address the issues of area limitations, diminishing the maintenance period in the hatchery and cost of production (Juinio-Meñez et al. 2012). Sea cucumber *H. scabra* has been cultured in sea ranching or sea farming and ponds in the nursery and the grow-out phase. It is necessary to create a cost-effective system to scale up the production of juveniles, particularly to grow them out to marketable size. Some studies focused on nursery production in land-based facilities such as Pitt and Nguyen (2004), who carried out the first mono species nursery trials using hapa nets or nursery cages in marine ponds and reared juveniles in hapa nets in a fiberglass hatchery tank in Vietnam. In the Philippines, nursery trials were conducted using hapa nets and pens in marine ponds and floating hapas in inland seawater channels (Juinio-Meñez et al. 2012; Gamboa et al. 2012). Purcell and Agudo (2013) demonstrated the use of mesh enclosures in the nursery seawater ponds in New Caledonia. A land-based nursery system in Indonesia utilized floating and fixed net cages (Firdaus and Indriana 2019) (Figure 4.8). Grow-out of *H. scabra* has been developing in earth pond and sea farming in Lombok, Indonesia (Figure 4.9). Monitoring growth performance of *H. scabra* juveniles is carried out every two weeks (Figure 4.10, Juinio-Meñez et al. 2017; Altamirano and Noran-Baylon 2020).

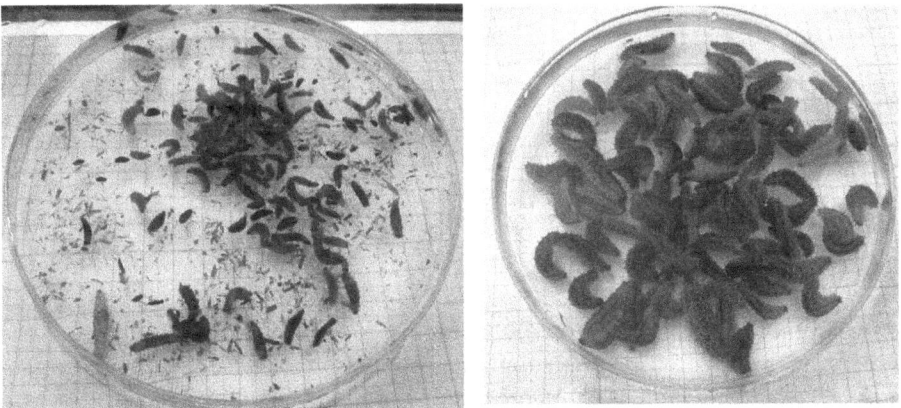

Figure 4.7 *Early juveniles of sea cucumber* Holothuria scabra.

Figure 4.8 *Nursery system of* Holothuria scabra *in earth pond in Lombok, Indonesia.*

Figure 4.9 *Grow-out of* Holothuria scabra *juveniles in earth pond and sea farming in Lombok Indonesia.*

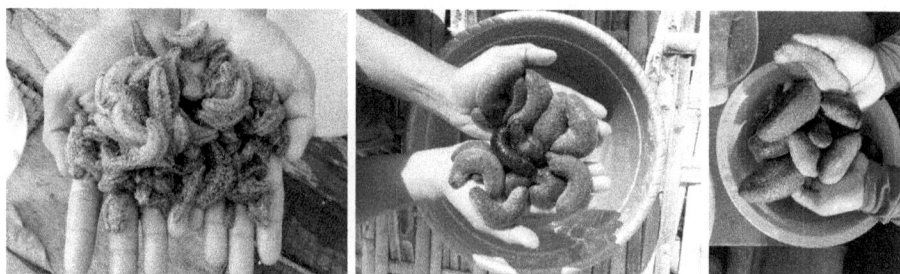

Figure 4.10 *Monitoring growth performance of* Holothuria scabra *juveniles cultured in pond and sea farming.*

Development of various grow-out systems provide solutions regarding the emerging problems in enhancing the production of sea cucumber until it reaches the marketable size. As reported by previous studies, land- and ocean-based grow-out systems have been well established and developed in ponds and the sea using pens, cages and also the sea ranching method (Pitt and

Nguyen 2004; Duy 2012; Juinio-Meñez et al. 2013; Juinio-Meñez, Evangelio, and Miralao 2014; Juinio-Meñez et al. 2017).

4.4.1.5 Co-Culture

The co-culture of sea cucumber with another animal becomes feasible to perform through a system called IMTA (Integrated Multi-Trophic Aquaculture). The IMTA system can provide advantages for both environmental, particularly in sustainability aquaculture, and social economic aspects (Chary et al. 2020). Sea cucumbers as deposit feeders consuming the particulate waste of other organisms are the best candidate for the IMTA system (Slater and Carton 2007).

An integrated mariculture system of *H. scabra* and seaweed *Eucheuma denticulatum* has been well-conducted in Tanzania (Namukose et al. 2016). In the same country, *H. scabra* is also reared with red seaweed *Kappaphycus striatum* (Beltranz-Guiterres et al. 2014). Purcell, Patrois and Fraisse (2006) investigated the co-culture of *H. scabra* with blue shrimp *Litopenaeus stylirostris* and reported its possibility, particularly in the early juvenile phase. However, it is not suggested to grow out both *H. scabra* and shrimp *L. stylirostris* in ponds (Bell et al. 2007). Sea cucumber *A. mollis* has a high survival rate when it is cultured with green-lipped mussel farms *Perna canaliculus* (Slater and Carton 2007). Furthermore, *H. scabra* has potential of bioremediation when it is co-cultured with red drum *Sciaenops ocellatus* in an IMTA system (Chary et al. 2020). Kim et al. (2015) investigated that in the co-culture cage systems, juvenile sea cucumber *A. japonicus* fed on the biodeposits of juvenile abalone, *Haliotis discus hannai*, reared in the same place.

4.4.2 Starfish

4.4.2.1 Animal Management

In order to produce good quality of larvae, adult starfish were collected in the peak spawning season, and this condition can be predicted based on analysis of relative gonad sizes (Keesing et al. 1997; Murabe et al. 2021; Balogh and Byrne 2020). Starfish were selected by observation of female and male traits, which include orange-colored ovaries and white-colored testes respectively (Murabe et al. 2021). Captured animals were placed into cool boxes during transportation (Balogh and Byrne 2020). In the laboratory, animals were reared in a circulating seawater tank (Keesing et al. 1997; Murabe et al. 2021) with biofilm-covered rocks as a diet source for *Parvulastra exigua* (Balogh and Byrne 2020). In another situation, *Asterias rubens* were fed on mussels in a holding tank (Agüera, Jansen, and Smaal 2020). Wilmes et al. (2018) determined that starfish *Acanthaster* sp. provided with coral *Acropora* sp. has a bigger gonad than with coral *Porites* sp. as diet. Davydov, Shubravyi and Vassetzky (1990) and Murabe et al. (2021) recommended macroalgae (i.e. *Codium fragile*, *C. itricatum*, *Ulva* sp. and *Porphyra* sp.), microphytes, fish, mollusks, polychaetes, as diet for adult *Asterina pectinifera*. For maintenance in the laboratory, water quality parameters such as salinity, temperature, pH and nitrate were observed weekly (Murabe et al. 2021).

Several studies observed starfish in certain species as pests. Starfish *A. rubens* is a predator of the blue mussel *Mytilus edulis* (Saier 2001; Kamermans, Blankendaal, and Perdon 2009; Agüera, Jansen, and Smaal 2020). Wilcox and Jeffs (2019) studied the mortality of green-lipped mussels, *Perna canaliculus*, caused by the predation of starfish *Coscinasterias muricata*.

4.4.2.2 Spawning Induction and Fertilization

Starfish culture has been developed in large-scale laboratory cultures. The culturing stages of sea starfish are spawning, fertilization, larvae rearing, settlement and grow-out (Keesing et al. 1997). Various forms of spawning induction have been developed to stimulate starfish, for instance *Marthasteria glacialis*, *Astropecten aurantiacus*, *Ceramaster placenta* and *P. exigua* injected with 10^{-5} M solution of 1-methyladenine into the coelom (Kanatani 1969; Balogh and Byrne 2020). In addition, Keesing et al. (1997) induced *Acanthaster planci* by injections of 1 ml of 1-methyladenine (10^{-3} M solution) into each arm above the gonad. *Asterias amurensis* was triggered by thermal shock (Kashenko 2005b). *A. pectinifera* was induced by hormonal stimulation (Davydov, Shubravyi, and Vassetzky 1990). The ripest males and females were selected by biopsy with the hormonal stimulation inserted into the arm (Keesing et al. 1997).

The fertilized egg cells were transferred into the parabolic larvae culture tanks with a single point air source and filled with filtered sea water (Keesing et al. 1997). The blastula and gastrulation processes were reached after fertilization occurred. When dipleurula becomes early bipinnaria, the digestive system, including the esophagus, stomach and intestine, appear (Kashenko 2005b). Long slender pigmented arms appeared in the late bipinnaria to early brachiolaria (Kashenko 2005b).

4.4.2.3 Larvae Rearing and Settlement

According to Kashenko (2005a) the developmental stages of *A. pectinifera* post fertilization are medium blastula, followed by ciliated blastula rotating in egg envelope, then the release of the blastula; in this condition larvae start to swim. After gastrula, the next stages reached are early dipleurula (without opening of the mouth) and dipleurula (with opening of the mouth). Following this, the stage moves to early bipinnaria and bipinnaria; next the stage enters into early brachiolaria, then brachiolaria with primordia of radial canals of the adult starfish and late brachiolaria (prior to settlement). Finally, the larvae metamorphose to juvenile.

Growth rate, survival, form, development and period of the planktonic stage of the larvae were affected not only by quality and size of egg but also the larvae diet (George 1999). Larvae were selected by criteria of active swimming and health; furthermore, poor quality larvae were discarded (Keesing et al. 1997). During the larvae rearing process, the density is a crucial factor of success in maintaining Echinodermata larvae. High density is more susceptible to contamination by bacteria and protozoa, and growth and development rates become relatively slow (Wray, Kitazawa, and Miner 2004). Water exchange is necessary to minimize the risk of contamination (Wray, Kitazawa, and Miner 2004). Total water replacement began on the fourth day after fertilization and continued every two to three days (Keesing et al. 1997; Kashenko 2005b).

Feeding is given when the digestive tract has formed completely and entered into the bipinnaria stage (Murabe et al. 2021). Microalga feeding regimes were provided approximately 5,000–10,000 cell ml^{-1} for larvae rearing including *Dunaliella tertiolecta* and *Phaeodactylum tricornutum* for *A. Planci*; *Rhodomonas* sp. for *Pisaster ochraceus*; *Chaetceros gracilis* and *C. Calcitrans* for *A. pectinifera*; *Nannochloris maculate*, *Isochrysis galbana* and *C. muelleri* for *A. amurensis* (Keesing et al. 1997; Okaji, Ayukai, and Lucas 1997; George 1999; Kashenko 2005b; Murabe et al. 2021). Furthermore, during *A. pectinifera* cultivation, the dipleurula stage were provided small microalgae (i.e. *N. maculata, Isochrysis galbana* and *C. muelleri*); moreover bipinnaria were fed large microalgae (i.e. *Phaeodactylum tricornutum* and *D. salina*) (Kashenko 2005a).

Bipinnaria larvae developed into brachiolaria larvae with the presence of three arms at the anterior end (Murabe et al. 2021). When larvae reached late brachiolaria, settlement substrate was necessary. Keesing et al. (1997) determined thalli of the coralline algae *Lithothumnion* sp. to induce settlement and metamorphosis *A. planci*. Prior to settlement, *A. amurensis* larvae were swimming to the bottom and attached into substrate which was covered with bacterial film (Kashenko 2005b). Davydov, Shubravyi and Vassetzky (1990) stimulated settling of *A. pectinifera* with mussel or oyster shells coated with a bacterial-algal film. Late brachiolariae and settled juvenile of *A. pectinifera* were fed dead cells of the microalgae (Kashenko 2005a). During the settlement phase, larvae were reared in the tank with gentle aeration and seawater was replaced each day.

4.4.2.4 Grow-Out of Starfish

In the planktonic phase brachiolaria metamorphoses into the benthic adult (Murabe et al. 2021). During the cultivation, settled juveniles were placed into grow-out tanks with slow circulation water (Keesing et al. 1997). In addition, densities of juveniles and adult *Acanthaster* sp. were affected by the survival and transition process in pre-settlement, settlement and post-settlement phases (Wilmes et al. 2018). Environmental factors influenced feeding activity, in particular change temperature, light intensity and salinity (Chen and Chen 1993; Agüera, Jansen, and Smaal 2020). Keesing et al. (1997) found that early juveniles of *A. planci* were fed with coralline algae, then juveniles 2–3 mm in length were provided diet of coral rubble collected from the reef as natural substrates. Finally, live coral, mussel and scallop meat were added as feeding when juveniles reached 10–15 mm and continued to gain bigger size. Furthermore, post-metamorphosed juvenile *Patiriella pseudoexigua* were provided benthic microalgae *Navicula* sp. on a diet (Chen and Chen 1993).

4.4.3 Sea Urchin

4.4.3.1 Broodstock Conditioning

Adult sea urchins are collected from the wild during the natural reproductive season by diving or snorkeling (Paredes, Bellas, and Costas 2015; Sonnenholzner-Varas, Touron, and Orrala 2018; Neves, Contins, and Nascimento 2018). Certain species can be found in shallow waters. Individuals are put in double plastic bags filled with seawater and the bags are packed in styrofoam containers to avoid direct contact with the light in order to keep the pre-spawning state (Sonnenholzner-Varas, Touron, and Orrala 2018).

The sea urchins are gently rinsed and are reared in outdoor flow-through seawater tanks containing sand-filtered and UV-treated seawater. Number of studies have been conducted to determine the diet regime of sea urchins during the holding period. Macroalgae such as *Ulva lactuca, Padina durvillaei, Laminaria digitate, L. saccharina* and *Palmaria palmata* have been introduced as the diet for sea urchins prior to spawning (Kelly et al. 2000; Dworjanyn and Pirozzi 2008; Paredes, Bellas, and Costas 2015; Sonnenholzner-Varas, Touron, and Orrala 2018; Suckling, Terrey, and Davies 2018). Kelp (*Saccharina japonica*) and the sporophyll of fresh *Undaria pinnatifida* as diet are reported to increase the gonad quality of sea urchin *Mesocentrotus nudus* including gonad size, color, hardness, texture and consistency (Takagi, Murata, and Agatsuma 2020; Takagi et al. 2017). Cirino, Ciaravolo et al. (2017) created a Ration Block of Food as an effective diet formulation of *Paracentrotus lividus* containing mussel meal, corn, natural macro-algae *U. lactuca* and microalgae *Spirulina platensis*, fish oil as well as calcium carbonate. In addition, *P. Lividus* were fed *ad libitum* on the red macroalgae *Palmaria palmate* (Castilla-Gavilán et al. 2018).

4.4.3.2 Spawning Induction and Fertilization

Different methods of spawning induction in sea urchin species have been reported by many previous studies. Sonnenholzner-Varas, Touron and Orrala (2018) induced the spawning of *Tripneustes depressus* by doing manual shaking and temperature shock up to 4–5 °C for 10 min. Another induction method was performed on *T. gratilla* by carrying out intracoelomic injection of 1 ml of 2 M potassium chloride (Dworjanyn and Pirozzi 2008) as well as on *Psammechinus miliaris* and *P. lividus*, stimulated by injecting 0.5–1 mL of

0.5 M KCl into the haemocoel (Kelly et al. 2000; Suckling, Terrey, and Davies 2018; Castilla-Gavilán et al. 2018). Cirino, Ciaravolo et al. (2017) induced the spawning of *P. Lividus* by manual shaking and electrical shock at low voltage. Spawning of *Centrostephanus rodgersii* is stimulated by 1–5 mL intracoelomic injection of 1.0 M KCl (Mos, Byrne, and Dworjanyn 2020).

The male and the female gametes are collected and are blended for the fertilization. Once the fertilization completes, the eggs are washed and are rinsed gently by passing them into a sieve to remove excess sperm. Embryos are placed into a holding tank containing UV-sterilized seawater with mild aeration (Dworjanyn and Pirozzi 2008; Mos, Byrne, and Dworjanyn 2020). Seawater in the tank is replaced twice a day (Mos, Byrne, and Dworjanyn 2020).

4.4.3.3 Embryonic Development and Larviculture

Embryonic development of sea urchins involves blastula, early gastrula, gastrula, prismatic larvae, early pluteus, four-arm pluteus, six-arm pluteus and eight-arm pluteus, etc. Larvae are maintained in aerated tanks containing seawater and the water is changed with 50–100% total volume every week (Paredes, Bellas, and Costas 2015; Sonnenholzner-Varas, Touron, and Orrala 2018; Suckling, Terrey, and Davies 2018). The bottom of the tank is siphoned daily (Liu and Chang 2015). Water quality parameters such as temperature and salinity are monitored every day while nitrate is measured before and after the water exchange is performed (Suckling, Terrey, and Davies 2018).

Three-days old larvae can be fed (Dworjanyn and Pirozzi 2008; Mos, Byrne, and Dworjanyn 2020) with single or mixed unicellular microalgae such as *Tetraselmis suecica, Isochrysis galbana, Chaetoceros gracilis, C. muelleri, Phaeodactilum tricornutum, Cylindrotheca closterium, Dunaliella tertiolecta, Tysochrisis lutea, Proteomonas sulcata* and *Rhodomonas* sp. (Dworjanyn and Pirozzi 2008; Paredes, Bellas, and Costas 2015; Brundu et al. 2016; Sonnenholzner-Varas, Touron, and Orrala 2018; Suckling, Terrey, and Davies 2018; Castilla-Gavilán et al. 2018; Mos, Byrne, and Dworjanyn 2020).

4.4.3.4 Settlement and Juvenile Culture

Various settlement plates and methods have been extensively studied. At the settlement, larvae are reared in tanks containing filtered seawater then plates are set as source of supplementary food. Water exchange and siphoning the tank bottom are important steps to remove harmful organisms to ensure the survival rate of larvae. Prior to settlement, plastic wave plates placed in the tank are conditioned with benthic diatom *Navicula* sp., *Cocconeis* sp. and *Nitzschia* (Liu and Chang 2015). Larvae, then, can be transferred to the settlement tank, pre-conditioned with an inoculum of the benthic diatom *Navicula* sp. at the bottom to develop biofilm as an inducer and source of feeding (Sonnenholzner-Varas, Touron, and Orrala 2018). Dworjanyn and Pirozzi

(2008) studied that macroalgae such as brown algae (i.e. *Dictyota dichotoma, Dilophus marginatus, Ecklonia radiata, Homeostrichus olsenii Womersley, Sargassum linearifolium* and *Zonaria angustata*), red algae (i.e. *Laurencia obtusa, Laurencia rigida*) and green alga (i.e. *U. lactuca*) could induce the settlement of *T. gratilla*. Dworjanyn and Pirozzi (2008) recommended the utilization of macroalgae without eliminating surface bacteria. Brundu et al. (2016) reported that post larvae of sea urchin *P. lividus* used settlement substrate covered with natural biofilm of diatom and cultured *Ulvella lens*. Castilla-Gavilán et al. (2020) suggested the introduction of diatom *Nitzschia laevis* biofilms as inducers in the settlement of *P. Lividus*. Sonnenholzner-Varas, Touron and Orrala (2018) provided macroalgae *Cladophora* sp. and semi-dried small pieces of *Padina durvillaei* as feed at the post-settlement stage.

According to Liu and Chang (2015), when feed in the settlement plates is completely consumed, early juveniles are then transferred into nursery tank. In the nursery tank, late juveniles were cultured in mesh cages in a land-based flow-through system. In the juveniles stage, Hu et al. (2020) reported the potential use of macroalgae *Saccharina japonica* as feeding for *Strongylocentrotus intermedius*. The result showed that wet body weight, specific growth rate, lantern size, wet gut weight and gonad yield were higher than those on the *Gracilaria lemaneiformis* diet. However, *G. lemaneiformis* was recommended for accelerating gonadal development of *S. intermedius*. Liu and Chang (2015) reported that there were various methods applied in order to develop grow-out of sea urchins including land-based rearing systems using fiberglass tanks, longline cultures using cage nets, benthic cage rearing at the bottom and sea ranching.

4.4.3.5 Integrated Multi-Trophic Aquaculture

The IMTA system has the potential of long-term culture at the semi-commercial scale by co-culturing any species from multiple trophic levels. IMTA has a number of potential benefits, namely, decreasing negative environmental impact, bioremediation by declining of inorganic and organic inputs, and biocontrol for other organisms (Lodeiros and García 2004; Orr et al. 2014).

Sea urchin *Lytechinus variegatus* is proposed as biocontrol for controlling net and shell fouling in pearl oyster *Pinctada imbricate* culture (Lodeiros and García 2004). According to Shpigel et al. (2018), sea urchin *P. lividus* is suitable to be cultured with fish *Sparus aurata* and seaweed *U. lactuca*. This system is capable to develop three reproductive cycle per year, to increase growth and quality of gonads, to decrease the growth period, to enhance the waste management by *U. lactuca* as biofilter and the fish provide nutrients for *U. lactuca*.

Developing an IMTA system between Echinodermata species is also feasible. Grosso et al. (2020) studied the rearing of sea urchin *P. lividus* as primary species, and sea cucumber *Holothuria tubulosa* as extractive species in an IMTA system. The result was successful with a high survival and growth rate. Orr et al. (2014) studied that green sea urchins *Strongylocentrotus droebachiensis* was able to ingest and absorb organic material from the solid waste of sablefish *Anoplopoma fimbria*. This result indicates that those animals are suitable candidates for an IMTA system.

Conclusions

Echinodermata have promising aspects and great potential in nutraceuticals industries. Sustainable aquacultures for Echinodermata have been developed for some species; developments of other Echinodermata species are also required to maintain sustainability of this marine phylum. Further, innovation on Echinodermata-based nutraceuticals should be more focused on bioavailability, heavy metals limit, efficacy, clinical trials and also quality improvement of currently available products.

References

Afzali, Mahbubeh, Javad Baharara, Khadijeh Nezhad Shahrokhabadi, and Elaheh Amini. 2017. "Evaluation of the cytotoxic effect of the brittle star (ophiocoma erinaceus) dichloromethane extract and doxorubicin on EL4 cell line." *Iranian Journal of Pharmaceutical Research: IJPR* 16 (Suppl):216.

Agudo, Natacha. 2006. *Sandfish hatchery techniques.* ACIAR; SPC. WorldFish Center.

Agüera, Antonio, Jeroen M. Jansen, and Aad C. Smaal. 2020. "Blue mussel (Mytilus edulis L.) association with conspecifics affects mussel size selection by the common seastar (Asterias rubens L.)." *Journal of Sea Research* 164:101935.

Altamirano, Jon P., and Roselyn D. Noran-Baylon. 2020. "Nursery culture of sandfish Holothuria scabra in sea-based floating hapa nets: Effects of initial stocking density, size grading and net replacement frequency." *Aquaculture* 526:735379.

Amini, Elaheh, Mohammad Nabiuni, Javad Baharara, Kazem Parivar, and Javad Asili. 2017. "In-vitro pro apoptotic effect of crude saponin from Ophiocoma erinaceus against cervical cancer." *Iranian Journal of Pharmaceutical Research: IJPR* 16 (1):266.

Arafa, S., M. Chouaibi, S. Sadok, and A. El Abed. 2012. "The influence of season on the gonad index and biochemical composition of the sea urchin Paracentrotus lividus from the Golf of Tunis." *The Scientific World Journal* 2012:815935. http://doi.org/10.1100/2012/815935.

Archana, A., and K. R. Babu. 2016. "Nutrient composition and antioxidant activity of gonads of sea urchin Stomopneustes variolaris." *Food Chem* 197 (Pt A):597–602. http://doi.org/10.1016/j.foodchem.2015.11.003.

Ardiansyah, Ardi, Abdullah Rasyid, Evi Amelia Siahaan, Ratih Pangetistu, and Tutik Murniasih. 2020. "Nutritional value and heavy metals content of sea cucumber holothuria scabra commercially harvested in Indonesia." *Current Research in Nutrition and Food Science Journal* 8 (3):765–773.

Baharara, Javad, and Elaheh Amini. 2015. "The potential of brittle star extracted polysaccharide in promoting apoptosis via intrinsic signaling pathway." *Avicenna Journal of Medical Biotechnology* 7 (4):151.

Baharara, Javad, Elaheh Amini, and Marzieh Mousavi. 2015. "The anti-proliferative and anti-angiogenic effect of the methanol extract from brittle star." *Reports of Biochemistry & Molecular Biology* 3 (2):68.

Baharara, Javad, Elaheh Amini, and Marziyeh Musavi. 2017. "Anti-vasculogenic activity of a polysaccharide derived from brittle star via inhibition of VEGF, Paxillin and MMP-9." *Iranian Journal of Biotechnology* 15 (3):179.

Bahrami, Yadollah, Wei Zhang, and Chris Franco. 2014. "Discovery of novel saponins from the viscera of the sea cucumber holothuria lessoni." *Marine Drugs* 12 (5):2633–2667.

Balić, Anamaria, Domagoj Vlašić, Kristina Žužul, Branka Marinović, and Bukvić Mokos. 2020. "Omega-3 versus omega-6 polyunsaturated fatty acids in the prevention and treatment of inflammatory skin diseases." *International Journal of Molecular Sciences* 21 (3):741.

Balogh, Regina, and Maria Byrne. 2020. "Developing in a warming intertidal, negative carry over effects of heatwave conditions in development to the pentameral starfish in Parvulastra exigua." *Marine Environmental Research* 162:105083.

Battaglene, Stephen C., J. Evizel Seymour, Christian Ramofafia, and Idris Lane. 2002. "Spawning induction of three tropical sea cucumbers, Holothuria scabra, H. fuscogilva and Actinopyga mauritiana." *Aquaculture* 207 (1–2):29–47.

Bell, Johann D., Natacha N. Agudo, Steven W. Purcell, Pascal Blazer, Mateo Simutoga, Dominique Pham, and Luc Della Patrona. 2007. "Grow-out of sandfish Holothuria scabra in ponds shows that co-culture with shrimp Litopenaeus stylirostris is not viable." *Aquaculture* 273 (4):509–519.

Beltranz-Guiterres, Marisol, Sebastian C.A. Ferse, A. Kunzmann, S.M. Stead, Flower E. Msuya, Thomas S. Hoffmeister, and Matthew J. Slater. 2014. "Co-culture of sea cucumber Holothuria scabra and red Kappa-phycus striatum." *Aquaculture Research* 47 (5):1549–1559. http://doi.org/10.1111/are.12615.

Brasseur, Lola, Guillaume Caulier, Patrick Flammang, Pascal Gerbaux, and Igor Eeckhaut. 2018. "Mapping of spinochromes in the body of three tropical shallow water sea urchins." *Natural Product Communications* 13 (12):1934578X1801301222.

Brasseur, L., E. Hennebert, L. Fievez, G. Caulier, F. Bureau, L. Tafforeau, P. Flammang, P. Gerbaux, and I. Eeckhaut. 2017. "The roles of spinochromes in four shallow water tropical sea urchins and their potential as bioactive pharmacological agents." *Marine Drugs* 15 (6). http://doi.org/10.3390/md15060179.

Brundu, Gianni, Lorena Vian Monleón, Dario Vallainc, and Stefano Carboni. 2016. "Effects of larval diet and metamorphosis cue on survival and growth of sea urchin post-larvae (Paracentrotus lividus; Lamarck, 1816)." *Aquaculture* 465:265–271.

Carvalho, Cristina, Renato X. Santos, Susana Cardoso, Sónia Correia, Paulo J. Oliveira, Maria S. Santos, and Paula I. Moreira. 2009. "Doxorubicin: the good, the bad and the ugly effect." *Current Medicinal Chemistry* 16 (25):3267–3285.

Castilla-Gavilán, Marta, Florence Buzin, Bruno Cognie, Justine Dumay, Vincent Turpin, and Priscilla Decottignies. 2018. "Optimising microalgae diets in sea urchin Paracentrotus lividus larviculture to promote aquaculture diversification." *Aquaculture* 490:251–259.

Castilla-Gavilán, Marta, Meshi Reznicov, Vincent Turpin, Priscilla Decottignies, and Bruno Cognie. 2020. "Sea urchin recruitment: Effect of diatom based biofilms on Paracentrotus lividus competent larvae." *Aquaculture* 515:734559.

Chary, Killian, Joël Aubin, Bastien Sadoul, Annie Fiandrino, Denis Covès, and Myriam D. Callier. 2020. "Integrated multi-trophic aquaculture of red drum (Sciaenops ocellatus) and sea cucumber (Holothuria scabra): Assessing bioremediation and life-cycle impacts." *Aquaculture* 516:734621.

Chen, C.-P., and B.-Y. Chen. 1993. "The effect of temperature-salinity combinations on survival and growth of juvenile Patiriella pseudoexigua (Echinodermata: Asteroidea)." *Marine Biology* 115 (1):119–122.

Cirino, P., C. Brunet, M. Ciaravolo, C. Galasso, L. Musco, T. Vega Fernández, C. Sansone, and A. Toscano. 2017. "The sea urchin arbacia lixula: A novel natural source of astaxanthin." *Mar Drugs* 15 (6). http://doi.org/10.3390/md15060187.

Cirino, Paola, Martina Ciaravolo, Angela Paglialonga, and Alfonso Toscano. 2017. "Long-term maintenance of the sea urchin Paracentrotus lividus in culture." *Aquaculture Reports* 7:27–33.

Claereboudt, Emily J. S., Guillaume Caulier, Corentin Decroo, Emmanuel Colson, Pascal Gerbaux, Michel R. Claereboudt, Hubert Schaller, Patrick Flammang, Magali Deleu, and Igor Eeckhaut. 2019. "Triterpenoids in echinoderms: Fundamental differences in diversity and biosynthetic pathways." *Marine drugs* 17 (6):352. http://doi.org/10.3390/md17060352.

Czarkwiani, Anna, Cinzia Ferrario, David Viktor Dylus, Michela Sugni, and Paola Oliveri. 2016. "Skeletal regeneration in the brittle star Amphiura filiformis." *Frontiers in Zoology* 13 (1):1–17.

D'Angelo, Stefania, Maria Letizia Motti, and Rosaria Meccariello. 2020. "ω-3 and ω-6 Polyunsaturated fatty acids, obesity and cancer." *Nutrients* 12 (9):2751.

D'Auria, Maria Valeria, Luigi Gomez Paloma, Luigi Minale, Raffaele Riccio, and Angela Zampella. 1995. "On the composition of sulfated polyhydroxysteroids in some ophiuroids and the structure determination of six new constituents." *Journal of Natural Products* 58 (2):189–196.

Davydov, P.V., O.I. Shubravyi, and S.G. Vassetzky. 1990. "The starfish Asterina pectinifera." In *Animal species for developmental studies*, 287–311. Springer.

Dincer, T., and S. Cakli. 2007. "Chemical composition and biometrical measurements of the Turkish sea urchin (Paracentrotus lividus, Lamarck, 1816)." *Critical Reviews in Food Science and Nutrition* 47 (1):21–26.

Domínguez-Godino, Jorge A., and Mercedes González-Wangüemert. 2018. "Breeding and larval development of Holothuria mammata, a new target species for aquaculture." *Aquaculture Research* 49 (4):1430–1440.

Domínguez-Godino, Jorge A., Matthew J. Slater, Colin Hannon, and Mercedes González-Wangüermert. 2015. "A new species for sea cucumber ranching and aquaculture: Breeding and rearing of Holothuria arguinensis." *Aquaculture* 438:122–128.

Drazen, Jeffrey C., Charles F. Phleger, Michaela A. Guest, and Peter D. Nichols. 2008. "Lipid, sterols and fatty acid composition of abyssal holothurians and ophiuroids from the North-East Pacific Ocean: Food web implications." *Comparative Biochemistry and Physiology Part B: Biochemistry and Molecular Biology* 151 (1):79–87.

Duque, Carmenza, Jorge Rojas, Sven Zea, Alejandro J. Roccatagliata, Marta S. Maier, and Alicia M. Seldes. 1997. "Main sterols from the ophiuroids Ophiocoma echinata, Ophiocoma wendtii, Ophioplocus januarii and Ophionotus victoriae." *Biochemical Systematics and Ecology* 25 (8):775–778. http://doi.org/10.1016/S0305-1978(97)00068-9.

Duy, Nguyen D. Q. 2012. "Large-scale sandfish production from pond culture in Vietnam." *Asia–Pacific Tropical Sea Cucumber Aquaculture. ACIAR Proceedings* 136:34–39.

Dworjanyn, Symon A., and Igor Pirozzi. 2008. "Induction of settlement in the sea urchin Tripneustes gratilla by macroalgae, biofilms and conspecifics: A role for bacteria?" *Aquaculture* 274 (2–4):268–274.

Dydjow-Bendek, Dorota, and Pawel Zagoździon. 2020. "Total dietary fats, fatty acids, and omega-3/omega-6 ratio as risk factors of breast cancer in the polish population–a case-control study." *Vivo* 34 (1):423–431.

Fawzya, Yusro Nuri, Hedi Indra Januar, Rini Susilowati, and Ekowati Chasanah. 2015. "Chemical composition and fatty acid profile of some Indonesian sea cucumbers." *Squalen Bulletin of Marine and Fisheries Postharvest and Biotechnology* 10 (1):27–34.

Firdaus, M., and L. F. Indriana. 2019. "Nursery performance of sandfish holothuria scabra juveniles in tidal earthen pond using different types of cage." *IOP Conference Series: Earth and Environmental Science* 370 (2019):012024. http://doi.org/10.1088/1755-1315/370/1/012024.

Folmer, F., M. Jaspars, M. Dicato, and M. Diederich. 2010. "Photosynthetic marine organisms as a source of anticancer compounds." *Phytochemistry Reviews* 9 (4):557–579.

Galasso, C., C. Corinaldesi, and C. Sansone. 2017. "Carotenoids from marine organisms: Biological functions and industrial applications." *Antioxidants (Basel)* 6 (4). http://doi.org/10.3390/antiox6040096.

Gamboa, Ruth U., R. A. Aurelio, Daisy A. Ganad, Lance B. Concepcion, and Neil Angelo S. Abreo. 2012. "Small-scale hatcheries and simple technologies for sandfish (Holothuria scabra) production." *Asia–Pacific Tropical Sea Cucumber Aquaculture. ACIAR Proceedings* 136:63–74.

Gammone, Maria Alessandra, Eugenio Gemello, Graziano Riccioni, and Nicolantonio Orazio. 2014. "Marine bioactives and potential application in sports." *Marine Drugs* 12 (5):2357–2382.

Gammone, M. A., G. Riccioni, G. Parrinello, and N. D'Orazio. 2018. "Omega-3 Polyunsaturated fatty acids: Benefits and endpoints in sport." *Nutrients* 11 (1). http://doi.org/10.3390/nu11010046.

Garama, D., P. Bremer, and A. Carne. 2012. "Extraction and analysis of carotenoids from the New Zealand sea urchin Evechinus chloroticus gonads." *Acta Biochimica Polonica* 59 (1):83–5.

Gazha, Anna K., Lyudmila A. Ivanushko, Eleonora V. Levina, Sergey N. Fedorov, Tatyana S. Zaporozets, Valentin A. Stonik, and Nataliya N. Besednova. 2016. "Steroid sulfates from ophiuroids (Brittle Stars): Action on some factors of innate and adaptive immunity." *Natural Product Communications* 11 (6):749–752.

George, Sophie B. 1999. "Egg quality, larval growth and phenotypic plasticity in a forcipulate seastar." *Journal of Experimental Marine Biology Ecology* 237 (2):203–224.

Gianasi, Bruno L., Jean-François Hamel, and Annie Mercier. 2019. "Triggers of spawning and oocyte maturation in the commercial sea cucumber Cucumaria frondosa." *Aquaculture* 498:50–60.

Grosso, Luca, Arnold Rakaj, Alessandra Fianchini, Lorenzo Morroni, Stefano Cataudella, and Michele Scardi. 2020. "Integrated Multi-Trophic Aquaculture (IMTA) system combining the sea urchin Paracentrotus lividus, as primary species, and the sea cucumber Holothuria tubulosa as extractive species." *Aquaculture* 534:736268.

Gustafson, Kirk R., Naoya Oku, and Dennis J. Milanowski. 2004. "Antiviral marine natural products." *Current Medicinal Chemistry-Anti-Infective Agents* 3 (3):233–249.

Hou, Yakun, Alan Carne, Michelle McConnell, Adnan A. Bekhit, Sonya Mros, Kikuko Amagase, and Alaa El-Din A. Bekhit. 2020. "In vitro antioxidant and antimicrobial activities, and in vivo anti-inflammatory activity of crude and fractionated PHNQs from sea urchin (Evechinus chloroticus)." *Food Chemistry* 316:126339.

Hou, Yakun, Elena A. Vasileva, Alan Carne, Michelle McConnell, Alaa El-Din A. Bekhit, and Natalia P. Mishchenko. 2018. "Naphthoquinones of the spinochrome class: Occurrence, isolation, biosynthesis and biomedical applications." *RSC Advances* 8 (57):32637–32650.

Hu, Chaoqun, Haipeng Li, Jianjun Xia, Lvping Zhang, Peng Luo, Sigang Fan, Pengfei Peng, Haipeng Yang, and Jing Wen. 2013. "Spawning, larval development and juvenile growth of the sea cucumber Stichopus horrens." *Aquaculture* 404:47–54.

Hu, Chaoqun, Youhou Xu, Jing Wen, Lvping Zhang, Sigang Fan, and Ting Su. 2010. "Larval development and juvenile growth of the sea cucumber Stichopus sp.(Curry fish)." *Aquaculture* 300 (1–4):73–79.

Hu, Fangyuan, Jia Luo, Mingfang Yang, Jiangnan Sun, Xiaofei Leng, Mingtai Liu, Xiujin Liao, Jian Song, Yaqing Chang, and Chong Zhao. 2020. "Effects of macroalgae Gracilaria lemaneiformis and Saccharina japonica on growth and gonadal development of the sea urchin Strongylocentrotus intermedius: New insights into the aquaculture management in southern China." *Aquaculture Reports* 17:100399.

Ibrahim, Mona Y., Sheikheldin M. Elamin, Yousif B. Abu Gideiri, and Sayed M. Ali. 2015. "The proximate composition and the nutritional value of some sea cucumber species inhabiting the sudanese red sea." *Food Science and Technology Management* 41 (11–17).

Jiao, Heng, Xiaohui Shang, Qi Dong, Shuang Wang, Xiaoyu Liu, Heng Zheng, and Xiaoling Lu. 2015. "Polysaccharide constituents of three types of sea urchin shells and their anti-inflammatory activities." *Marine Drugs* 13 (9):5882–5900.

Juinio-Meñez, Marie Antonette, Julissah C. Evangelio, and Sasa James Miralao. 2014. "Trial grow-out culture of sea cucumber H olothuria scabra in sea cages and pens." *Aquaculture Research* 45 (8):1332–1340.

Juinio-Meñez, Marie Antonette, Julissah C. Evangelio, Ronald D. Olavides, Marie Antonette S. Paña, Glycinea M. De Peralta, Christine Mae A. Edullantes, Bryan Dave R. Rodriguez, and Inggat Laya N. Casilagan. 2013. "Population dynamics of cultured Holothuria scabra in a sea ranch: Implications for stock restoration." *Reviews in Fisheries Science* 21 (3–4):424–432.

Juinio-Meñez, Marie Antonette, Glycinea M. de Peralta, R.J.P. Dumalan, C.M. Edullantes, and T.O. Catbagan. 2012. "Ocean nursery systems for scaling up juvenile sandfish (Holothuria scabra) production: Ensuring opportunities for small fishers." *Asia–Pacific Tropical Sea Cucumber Aquaculture. ACIAR Proceedings* 136:57–62.

Juinio-Meñez, Marie Antonette, Elsie D. Tech, Isidora P. Ticao, Jay R.C. Gorospe, Christine Mae A. Edullantes, and Rose Angeli V. Rioja. 2017. "Adaptive and integrated culture production systems for the tropical sea cucumber Holothuria scabra." *Fisheries Research* 186:502–513.

Kamermans, Pauline, Monique Blankendaal, and Jack Perdon. 2009. "Predation of shore crabs (Carcinus maenas (L.)) and starfish (Asterias rubens L.) on blue mussel (Mytilus edulis L.) seed from wild sources and spat collectors." *Aquaculture* 290 (3–4):256–262.

Kanatani, H. 1969. "Induction of spawning and oocyte maturation by L-methyladenine in starfishes." *Experimental Cell Research* 57 (2–3):333–337.

Kashenko, S.D. 2005a. "Chronology of development in the starfish Asterina pectinifera from Vostok Bay, Sea of Japan." *Russian Journal of Marine Biology* 31 (4):261–264.

Kashenko, S.D. 2005b. "Development of the starfish Asterias amurensis under laboratory conditions." *Russian Journal of Marine Biology* 31 (1):36–42.

Keesing, John K., Andrew R. Halford, Karina C. Hall, and Carina M. Cartwright. 1997. "Large-scale laboratory culture of the crown-of-thorns starfish Acanthaster planci (L.) (Echinodermata: Asteroidea)." *Aquaculture* 157 (3–4):215–226.

Kelly, Maeve S., Amanda J. Hunter, Claire L. Scholfield, and J. Douglas McKenzie. 2000. "Morphology and survivorship of larval Psammechinus miliaris (Gmelin) (Echinodermata: Echinoidea) in response to varying food quantity and quality." *Aquaculture* 183 (3–4):223–240.

Kim, Se-Kwon, and S.W.A. Himaya. 2012. "Triterpene glycosides from sea cucumbers and their biological activities." *Advances in Food and Nutrition Research* 65:297–319.

Kim, Taeho, Ho-Seop Yoon, Seungsik Shin, Moo-Hwan Oh, Inyeong Kwon, Jihoon Lee, Sang-Duk Choi, and Kwan-Sik Jeong. 2015. "Physical and biological evaluation of co-culture cage systems for grow-out of juvenile abalone, Haliotis discus hannai, with juvenile sea cucumber, Apostichopus japonicus (Selenka), with CFD analysis and indoor seawater tanks." *Aquaculture* 447:86–101.

Klimenko, Antonina, Robin Huber, Laurence Marcourt, Estelle Chardonnens, Alexey Koval, Yuri S. Khotimchenko, Emerson Ferreira Queiroz, Jean-Luc Wolfender, and Vladimir L. Katanaev. 2021. "A cytotoxic porphyrin from North Pacific brittle star Ophiura sarsii." *Marine Drugs* 19 (1):11.

Laguerre, Hélène, Grégory Raymond, Patrick Plan, Nadia Améziane, Xavier Bailly, and Patrick Le Chevalier. 2020. "First description of embryonic and larval development, juvenile growth of the black sea-cucumber Holothuria forskali (Echinodermata: Holothuroidea), a new species for aquaculture in the north-eastern Atlantic." *Aquaculture* 521:734961.

Lee, Chi-Chiu, Hernyi Justin Hsieh, Cheng-Hong Hsieh, and Deng-Fwu Hwang. 2014. "Spine venom of crown-of-thorns starfish (Acanthaster planci) induces antiproliferation and apoptosis of human melanoma cells (A375. S2)." *Toxicon* 91:126–134.

Lee, Chi-Chiu, Hernyi Justin Hsieh, and Deng-Fwu Hwang. 2015. "Cytotoxic and apoptotic activities of the plancitoxin I from the venom of crown-of-thorns starfish (Acanthaster planci) on A375. S2 cells." *Journal of Applied Toxicology* 35 (4):407–417.

Li, Li, Qi Li, and Lingfeng Kong. 2010. "The effect of different substrates on larvae settlement in sea cucumber, Apostichopus japonicus Selenka." *Journal of the World Aquaculture Society* 41:123–130.

Liu, Hui, and Ya-Qing Chang. 2015. "Sea urchin aquaculture in China." *Echinoderm Aquaculture*:127–146.

Lodeiros, César, and Natividad García. 2004. "The use of sea urchins to control fouling during suspended culture of bivalves." *Aquaculture* 231 (1–4):293–298.

Lourenço, Sílvia, Luísa M.P. Valente, and Carlos Andrade. 2019. "Meta-analysis on nutrition studies modulating sea urchin roe growth, colour and taste." *Reviews in Aquaculture* 11 (3):766–781. https://doi.org/10.1111/raq.12256.

Mamelona, Jean, Richard Saint-Louis, and Émilien Pelletier. 2010. "Proximate composition and nutritional profile of by-products from green urchin and Atlantic sea cucumber processing plants." *Food Science Technology* 45 (10):2119–2126. https://doi.org/10.1111/j.1365-2621.2010.02381.x.

Mercier, Annie, Stephen C. Battaglene, and Jean-François Hamel. 2000. "Settlement preferences and early migration of the tropical sea cucumber Holothuria scabra." *Journal of Experimental Marine BiologyEcology* 249 (1):89–110.

Militz, Thane A., Esther Leini, Nguyen Dinh Quang Duy, and Paul C. Southgate. 2018. "Successful large-scale hatchery culture of sandfish (Holothuria scabra) using microalgae concentrates as a larval food source." *Aquaculture Reports* 9:25–30.

Mills, David J., Nguyen D.Q. Duy, Marie Antonette Juinio-Meñez, Christina M. Raison, and Jacques M. Zarate. 2012. "Overview of sea cucumber aquaculture and sea-ranching research in the South-East Asian region." *Asia–Pacific Tropical Sea Cucumber Aquaculture. ACIAR Proceedings*. Australian Centre for International Agricultural Research (ACIAR). https://hatcheryfm.com/article-files/file_1332369006_2.pdf

Mol, Suehendan, Tacnur Baygar, Candan Varlik, and Ş. Yasemin Tosun. 2008. "Seasonal variations in yield, fatty acids, amino acids and proximate compositions of sea urchin (Paracentrotus lividus) roe." 藥物食品分析 16 (2):68–74.

Mos, Benjamin, Maria Byrne, and Symon A. Dworjanyn. 2020. "Effects of low and high pH on sea urchin settlement, implications for the use of alkali to counter the impacts of acidification." *Aquaculture* 528:735618.

Murabe, Naoyuki, Ei-ichi Okumura, Kazuyoshi Chiba, Enako Hosoda, Susumu Ikegami, and Takeo Kishimoto. 2021. "The starfish asterina pectinifera: Collection and maintenance of adults and rearing and metamorphosis of larvae." In *Developmental biology of the sea urchin and other marine invertebrates*, 49–68. Springer.

Namukose, Mary, Flower E. Msuya, Sebastian C.A. Ferse, Matthew J. Slater, and Andreas Kunzmann. 2016. "Growth performance of the sea cucumber Holothuria scabra and the seaweed Eucheuma denticulatum: integrated mariculture and effects on sediment organic characteristics." *Aquaculture Environment Interactions* 8:179–189.

Neves, Raquel A.F., Mariana Contins, and Silvia M. Nascimento. 2018. "Effects of the toxic benthic dinoflagellate Ostreopsis cf. ovata on fertilization and early development of the sea urchin Lytechinus variegatus." *Marine Environmental Research* 135:11–17.

Nhu Hieu, Vo Mai, Tran Thi Thanh Van, Cao Thi Thuy Hang, Natalia P. Mischenko, Fedoreyev Sergey A., and Hai Bang Truong. 2020. "Polyhydroxynaphthoquinone pigment from Vietnam sea urchins as a potential bioactive ingredient in cosmeceuticals." *Natural Product Communications* 15 (11):1934578X20972525.

Okaji, K., T. Ayukai, and J.S. Lucas. 1997. "Selective feeding by larvae of the crown-of-thorns starfish, Acanthaster planci (L.)." *Coral Reefs* 16 (1):47–50.

Omran, Nahla El-Sayed El-Shazly. 2013. "Nutritional value of some Egyptian sea cucumbers." *African Journal of Biotechnology* 12 (35).

Orr, Lindsay C., Daniel L. Curtis, Stephen F. Cross, Helen Gurney-Smith, Alynn Shanks, and Christopher M. Pearce. 2014. "Ingestion rate, absorption efficiency, oxygen consumption, and fecal production in green sea urchins (Strongylocentrotus droebachiensis) fed waste from sablefish (Anoplopoma fimbria) culture." *Aquaculture* 422:184–192.

Pangestuti, Ratih, and Zainal Arifin. 2017. "Medicinal and health benefit effects of functional sea cucumbers." *Journal of Traditional and Complementary Medicine* 8 (3):341–351.

Pangestuti, Ratih, and Se-Kwon Kim. 2017. "Bioactive peptide of marine origin for the prevention and treatment of non-communicable diseases." *Marine drugs* 15 (67):1–23.

Pangestuti, Ratih, Evi Siahaan, and Se-Kwon Kim. 2018. "Photoprotective substances derived from marine algae." *Marine Drugs* 16 (11):399.

Paredes, E., J. Bellas, and D. Costas. 2015. "Sea urchin (Paracentrotus lividus) larval rearing—Culture from cryopreserved embryos." *Aquaculture* 437:366–369.

Patar, Azim, Hasnan Jaafar, Syed Mohsin Syed Sahil Jamalullail, and Jafri Malin Abdullah. 2012. "The body wall crude extract of Stichopus variegatus promotes repair of acute contused spinal cord injury in rats by improving motor function and reduces intramedullary hemorrhage." *Biomedical Research* 23 (4).

Patar, Azim, Syed Mohsin Syed Sahil Jamalullail, Hasnan Jaafar, and Jafri Malin Abdullah. 2012. "The effect of water extract of sea cucumber Stichopus variegatus on rat spinal astrocytes cell lines." *Current Neurobiology* 3 (1).

Pitt, Rayner, and D.Q.D. Nguyen. 2004. "Breeding and rearing of the sea cucumber Holothuria scabra in Viet Nam." *FAO Fisheries Technical Paper. No. 463.* Rome, FAO. 2004. 425p.

Purcell, Steven W., and Natacha S. Agudo. 2013. "Optimisation of mesh enclosures for nursery rearing of juvenile sea cucumbers." *PLoS ONE* 8 (5):e64103.

Purcell, Steven W., Bernard F. Blockmans, and Natacha N.S. Agudo. 2006. "Transportation methods for restocking of juvenile sea cucumber, Holothuria scabra." *Aquaculture* 251 (2–4):238–244.

Purcell, Steven W., Cathy A. Hair, and David Mills. 2012. "Sea cucumber culture, farming and sea ranching in the tropics: Progress, problems and opportunities." *Aquaculture* 368:68–81.

Purcell, Steven W., Jacques Patrois, and Nicolas Fraisse. 2006. "Experimental evaluation of co-culture of juvenile sea cucumbers, Holothuria scabra (Jaeger), with juvenile blue shrimp, Litopenaeus stylirostris (Stimpson)." *Aquaculture Research* 37 (5):515–522.

Rakaj, Arnold, Alessandra Fianchini, Paola Boncagni, Michele Scardi, and Stefano Cataudella. 2019. "Artificial reproduction of Holothuria polii: A new candidate for aquaculture." *Aquaculture* 498:444–453.

Rasyid, Abdullah, Tutik Murniasih, Masteria Y. Putra, Ratih Pangestuti, Iskandar A. Harahap, Febriana Untari, and Sari B.M. Sembiring. 2020. "Evaluation of nutritional value of sea cucumber Holothuria scabra cultured in Bali, Indonesia." *Aquaculture, Aquarium, Conservation & Legislation* 13 (4):2083–2093.

Ren, Yichao, Wenshan Liu, and Christopher M. Pearce. 2018. "Effects of stocking density, ration and temperature on growth, survival and metamorphosis of auricularia larvae of the California sea cucumber, Parastichopus californicus." *Aquaculture Research* 49 (1):517–525.

Riccio, Raffaele, Maria Valeria D'Auria, and Luigi Minale. 1985. "Unusual sulfated marine steroids from the ophiuroid ophioderma longicaudum." *Tetrahedron* 41 (24):6041–6046. http://doi.org/10.1016/S0040-4020(01)91445-0.

Roccatagliata, Alejandro J., Marta S. Maier, and Alicia M. Seldes. 1998. "New sulfated polyhydroxysteroids from the antarctic ophiuroid astrotoma a gassizii." *Journal of Natural Products* 61 (3):370–374.

Roccatagliata, Alejandro J., Marta S. Maier, Alicia M. Seldes, Carlos A. Pujol, and Elsa B. Damonte. 1996. "Antiviral sulfated steroids from the ophiuroid Ophioplocus januarii." *Journal of Natural Products* 59 (9):887–889.

Rocha, F., A. C. Rocha, L. F. Baião, J. Gadelha, C. Camacho, M. L. Carvalho, F. Arenas, A. Oliveira, M. R. G. Maia, A. R. Cabrita, M. Pintado, M. L. Nunes, C. M. R. Almeida, and L. M. P. Valente. 2019. "Seasonal effect in nutritional quality and safety of the wild sea urchin Paracentrotus lividus harvested in the European Atlantic shores." *Food Chem* 282:84–94. http://doi.org/10.1016/j.foodchem.2018.12.097.

Ruocco, N., S. Costantini, S. Guariniello, and M. Costantini. 2016. "Polysaccharides from the marine environment with pharmacological, cosmeceutical and nutraceutical potential." *Molecules* 21 (5). http://doi.org/10.3390/molecules21050551.

Saier, Bettina 2001. "Direct and indirect effects of seastars Asterias rubens on mussel beds (Mytilus edulis) in the Wadden Sea." *Journal of Sea Research* 46 (1):29–42.

Salem, Yosra Ben, Safa Amri, Khaoula Mkadmini Hammi, Amal Abdelhamid, Didier Le Cerf, Abderrahman Bouraoui, and Hatem Majdoub. 2017. "Physico-chemical characterization and pharmacological activities of sulfated polysaccharide from sea urchin, Paracentrotus lividus." *International Journal of Biological Macromolecules* 97:8–15.

Schulze, Matthias B., Anne Marie Minihane, Rasha Noureldin M. Saleh. 2020. "Intake and metabolism of omega-3 and omega-6 polyunsaturated fatty acids: nutritional implications for cardiometabolic diseases." *The Lancet Diabetes Risérus, and Endocrinology* 8 (11):915–930.

Shang, W. H., J. N. Yan, Y. N. Du, X. F. Cui, S. Y. Su, J. R. Han, Y. S. Xu, C. F. Xue, T. T. Zhang, H. T. Wu, and B. W. Zhu. 2020. "Functional properties of gonad protein isolates from three species of sea urchin: A comparative study." *Journal of Food Science* 85 (11):3679–3689. http://doi.org/10.1111/1750-3841.15464.

Shpigel, M., L. Shauli, V. Odintsov, D. Ben-Ezra, A. Neori, and L. Guttman. 2018. "The sea urchin, Paracentrotus lividus, in an Integrated Multi-Trophic Aquaculture (IMTA) system with fish (Sparus aurata) and seaweed (Ulva lactuca): Nitrogen partitioning and proportional configurations." *Aquaculture* 490:260–269.

Slater, Matthew J., and Alexander G. Carton. 2007. "Survivorship and growth of the sea cucumber Australostichopus (Stichopus) mollis (Hutton 1872) in polyculture trials with green-lipped mussel farms." *Aquaculture* 272 (1–4):389–398.

Sonnenholzner-Varas, Jorge I., Noelia Touron, and María Manuela Panchana Orrala. 2018. "Breeding, larval development, and growth of juveniles of the edible sea urchin Tripneustes depressus: A new target species for aquaculture in Ecuador." *Aquaculture* 496:134–145.

Suckling, Coleen C., Maeve S. Kelly, and Rachael C. Symonds. 2020. "Carotenoids in sea urchins." In *Developments in aquaculture and fisheries science*, 209–217. Elsevier.

Suckling, Coleen C., David Terrey, and Andrew J. Davies. 2018. "Optimising stocking density for the commercial cultivation of sea urchin larvae." *Aquaculture* 488:96–104.

Sun, Jenny, and Fu-Sung Chiang. 2015. "Use and exploitation of sea urchins." In Brown, N.P., and Eddy, S.D. (Eds.), *Echinoderm aquaculture*, 25–45. Wiley-Blackwell.

Svetashev, Vasily I., and Vladimir I. Kharlamenko. 2020. "Fatty acids of abyssal echinodermata, the sea star eremicaster vicinus and the sea urchin kamptosoma abyssale: A new polyunsaturated fatty acid detected, 22:6(n-2)." *Lipids* 55 (3):291–296. https://doi.org/10.1002/lipd.12227.

Swain, Shasank S., Rabindra N. Padhy, and Pawan K. Singh. 2015. "Anticancer compounds from cyanobacterium Lyngbya species: A review." *Antonie van Leeuwenhoek* 108 (2):223–265.

Symonds, R. C., M. S. Kelly, C. Caris-Veyrat, and A. J. Young. 2007. "Carotenoids in the sea urchin Paracentrotus lividus: occurrence of 9'-cis-echinenone as the dominant carotenoid in gonad colour determination." *Comparative Biochemistry and Physiology - B Biochemistry and Molecular Biology* 148 (4):432–444. http://doi.org/10.1016/j.cbpb.2007.07.012.

Takagi, Satomi, Yuko Murata, and Yukio Agatsuma. 2020. "Feeding the sporophyll of Undaria pinnatifida kelp shortens the culture duration for the production of high-quality gonads of Mesocentrotus nudus sea urchins from a barren." *Aquaculture* 528:735503.

Takagi, Satomi, Yuko Murata, Eri Inomata, Hikaru Endo, Masakazu N. Aoki, and Yukio Agatsuma. 2017. "Improvement of gonad quality of the sea urchin Mesocentrotus nudus fed the kelp Saccharina japonica during offshore cage culture." *Aquaculture* 477:50–61.

Thao, Nguyen Phuong, Nguyen Xuan Cuong, Bui Thi Thuy Luyen, Nguyen Van Thanh, Nguyen Xuan Nhiem, Young-Sang Koh, Bui Minh Ly, Nguyen Hoai Nam, Phan Van Kiem, and Chau Van Minh. 2013. "Anti-inflammatory asterosaponins from the starfish *Astropecten monacanthus.*" *Journal of Natural Products* 76 (9):1764–1770.

Ueoka, Reiko, Yuki Hitora, Akihiro Ito, Minoru Yoshida, Shigeru Okada, Kentaro Takada, and Shigeki Matsunaga. 2016. "Curacin E from the brittle star ophiocoma scolopendrina." *Journal of Natural Products* 79 (10):2754–2757.

Vasileva, Elena A., Natalia P. Mishchenko, Van T.T. Tran, Hieu Vo, and Sergey A. Fedoreyev. 2021. "Spinochrome identification and quantification in pacific sea urchin shells, coelomic fluid and eggs using HPLC-DAD-MS." *Marine Drugs* 19 (1):21.

Wang, Wei-Hong, Jong-Ki Hong, Chong-Ok Lee, Hee-Young Cho, Sook Shin, and Jee-H. Jung. 2004. "Bioactive Metabolites from the brittle star Ophioplocus japonicus." *Natural Product Sciences* 10 (6):253–261.

Wen, Jing, Chaoqun Hu, and Sigang Fan. 2010. "Chemical composition and nutritional quality of sea cucumbers." *Journal of the Science of Food and Agriculture* 90 (14):2469–2474.

Wilcox, Mark, and Andrew Jeffs. 2019. "Impacts of sea star predation on mussel bed restoration." *Restoration Ecology* 27 (1):189–197.

Wilmes, Jennifer C., Ciemon F. Caballes, Zara-Louise Cowan, Andrew S. Hoey, Bethan J. Lang, Vanessa Messmer, and Morgan S. Pratchett. 2018. "Contributions of pre-versus post-settlement processes to fluctuating abundance of crown-of-thorns starfishes (Acanthaster spp.)." *Marine Pollution Bulletin* 135:332–345.

Wray, Gregory A., Chisato Kitazawa, and Benjamin Miner. 2004. "Culture of echinoderm larvae through metamorphosis." *Methods in Cell Biology* 74:75–86.

Yang, Jingfeng, Meng Yi, Jinfeng Pan, Jun Zhao, Liming Sun, Xinping Lin, Yuegang Cao, Lu Huang, Beiwei Zhu, and Chenxu Yu. 2015. "Sea urchin (Strongylocentrotus intermedius) polysaccharide enhanced BMP-2 induced osteogenic differentiation and its structural analysis." *Journal of Functional Foods* 14:519–528.

Yu, Xiao-Qing, Chang-Sheng Jiang, Yi Zhang, Pan Sun, Tibor Kurtán, Attila Mándi, Xiao-Lu Li, Li-Gong Yao, Ai-Hong Liu, Bin Wang, Yue-Wei Guo, and Shui-Chun Mao. 2017. "Compositacins A–K: Bioactive chamigrane-type halosesquiterpenoids from the red alga Laurencia composita Yamada." *Phytochemistry* 136:81–93. https://doi.org/10.1016/j.phytochem.2017.01.007.

Yuan, Xiutang, Hongsheng Yang, Lili Wang, Yi Zhou, and Howaida Gabr. 2010. "Effects of salinity on energy budget in pond-cultured sea cucumber Apostichopus japonicus (Selenka)(Echinodermata: Holothuroidea)." *Aquaculture* 306 (1–4):348–351.

5

An Overview of the Secondary Uses of Zooplankton for Nutrients in the Food Chain

Khushali M. Pandya

Contents

5.1 Introduction

Recognizing the significant role of marine plankton in the food chain, there is a growing awareness of the unreleased potential found at the base of the food chain (Josué et al., 2019). The most important observation here is the species of zooplankton. Several species of zooplankton differ in their nutritive value and thus the food conversion ratio to enrichment in essential fatty acids or proteins differs (Ramlee et al., 2021). The equilibrium between phytoplankton, zooplankton, beneficial bacteria (pond probiotics), and other secondary consumers during the culture period of any organism plays a crucial role in the maintenance of aquatic organisms' health (Raman & Gopakumar, 2018). The secondary consumers—the zooplankton—occupy a crucial position in the food web of any ecosystem as they modulate the energy available to the following trophic levels. Another important characteristic of zooplankton is that they are directly affected by environmental changes. Zooplankton form one of the most important candidates for ecological indicators. Copepods' diets are now forming an implication of the traditional concept for the classical food chain. Copepods enriched with rice bran produced a high result of protein, lipids and are comparable with the results of copepods enriched with *Chlorella* sp. The use of agro-industrial residue such as rice bran is used to substitute microalgae culture for boosting the production of live feed and fish larvae. Enrichment of a mixed diet (carrots and spinach) has produced high

DOI: 10.1201/9781003128175-5

results of specific growth rate, survival rate and better enhancement of coloration on Betta species. An enrichment medium for copepods was developed to improve the growth performance, survival and coloration of *B. splendens*, which is important in the Betta trade industry (Yuslan et al., 2021). Large heterotrophic protists in the diet of copepods are an intermediate link between the plankton loop and higher trophic levels (Sherr & Sherr, 2002). Zooplankton can rapidly colonize and reproduce at a higher rate. The amino acid composition of the species is a good indicator of the trophic niche and the adaptations of the species to abiotic factors. Copepods are one major zooplankton group in marine systems in which daphniids are almost absent, whereas in freshwater systems daphniids are keystone zooplankter. Zooplankton acts as an important link between the primary producers (phytoplankton) and organisms on higher trophic levels, including marine fish and marine mammals.

5.2 Plankton Community

Decades of studies have indicated the importance of digestion of phytoplankton by zooplankton being the largest sink in the global marine carbon budget; the dynamicity in the adaptive capabilities of the aquatic organisms concerning the changing environmental conditions is exhibited in the form of physiological states. Any change in abiotic factors—viz., hydrodynamic conditions, temperature, and salinity along with food as a factor—affect the biochemical and metabolic processes strongly and respond via stress, as studied by various scientists. Previously it has been reported that a vital role is played by the amino acids during salinity acclimation. In recent years, special emphasis has been given to krill oil for human consumption from a nutraceutical perspective. Antarctic krill oil's composition has fatty acid composition comparable to fish oils. However, large proportions of the fatty acids are esterified (Tou et al., 2007). The fatty acid content of eicosapentaenoic acid (EPA) and docosahexaenoic acid (DHA) is rich. The population dynamics of zooplankton are largely dependent on food availability. The nutritive state of the zooplankton is indicated by the protein content in the subsequent trophic levels. Ketocarotenoid astaxanthin is also a compound accumulating in zooplankton and crustaceans which is produced by phytoplankton. The plankton literature suggests high food-quality algae species are rich in HUFA and low food-quality algae are poor in HUFA. Adding semi-pure emulsions of HUFA to algae monocultures can markedly increase the growth rates of zooplankton feeding on these mixtures. A study measuring zooplankton biomass accrual when feeding on natural phytoplankton found a strong correlation between phytoplankton HUFA (specifically eicosapentaenoic acid) content and herbivorous zooplankton production. The aquatic ecology literature suggests that planktonic food webs with high HUFA content phytoplankton have high zooplankton to phytoplankton biomass ratios, while systems with low HUFA phytoplankton have low zooplankton biomass. Also, the seasonal succession of plankton in many temperate lakes follows patterns tied to phytoplankton HUFA content, with intense zooplankton grazing and 'clear-waterphases' characteristic of periods when the phytoplankton is dominated by HUFA-rich species. Herbivorous zooplankton production is constrained by the zooplankton's ability to ingest and digest phytoplankton. It is becoming increasingly clear, however, that much of the phytoplankton which is assimilated may be nutritionally inadequate. HUFA may be key nutritional constituents of zooplankton diets, and may determine energetic efficiency across the plant-animal interface, secondary production and the strength of trophic coupling in aquatic pelagic food webs.

A biochemical fingerprint has been demonstrated in zooplankton species (Boëchat & Adrian, 2005). A discriminant analysis was performed at altitudinal variation in the amino acid composition of various zooplankton belonging to phylum Arthropoda—calanoids, cladocerans, cyclopoids, and rotifers. Leibold (1995) showed the composition of amino acids in zooplankton species utilized for food resource partitioning. The amino acid composition of marine copepods is species-specific and remains constant as per the studies conducted by Guisande et al. (2002). Fishes are unable to produce DHA by themselves through specific metabolic pathways utilizing the zooplankton. Accumulation of omega-3 fatty acids is achieved by feeding on zooplankton. The use of eicosapentaenoic acid (EPA) and DHA as feed supplements are increasing from an aquaculture perspective because of this secondary metabolism. Cowie and Hedges (1996) studied the digestion in herbivorous zooplankton which feeds upon the phytoplankton. They showed the quantitative estimation of the fate of aldoses and amino acids. Organic carbon and nitrogen are derived from aldoses and particularly amino acids as important sources to the copepods. The copepods have been able to convert most of the organic materials from the diatoms fed to them. These fecal pellets from the zooplankton are trapped in the sediment and collected in the environment (Wakeham & Lee, 1993). Selective digestion of components like ribose studied as a biomarker by Cowie and Hedges (1984) also acts as a quantitative tracer for the individual organic matter. Prahl et al. (1988) and Harvey et al. (1987) studied the correlations amongst dietary algal lipids such as fatty acids and alcohol and their digestion patterns which show the difference between pellet and diet. Zooplankton, such as copepods and krill, are the most numerous primary consumers of plankton in the marine environment (Figure 5.1, Gasmi et al., 2020). The copepod Calanus *finmarchicus* has lipid-rich stages for harvesting and its oil can be used as a health-promoting nutraceutical (Gasmi et al., 2020). These data present an opportunity to find out new sources of UFA-rich oil from the marine ecosystem. In recent times, C. *finmarchicus* is utilized as an alternative to marine fishes because of the presence of PUFA-rich oil in this zooplankton (Figure 5.2). C. *finmarchicus* acts as a critical link between the phytoplankton and other marine organisms present at the higher trophic levels in a marine food web. It contains a very high quantity of long-chain ω-3EPA and DHA (Pedersen et al., 2014). The oil extracted from C. *finmarchicus* is viscous owing to the presence of fatty acids in the wax ester form that comprises 80–90% of the oil composition.

Figure 5.1 *Zooplankton* Calanus finmarchicus.

Source: Cameron Thompson, the University of Maine

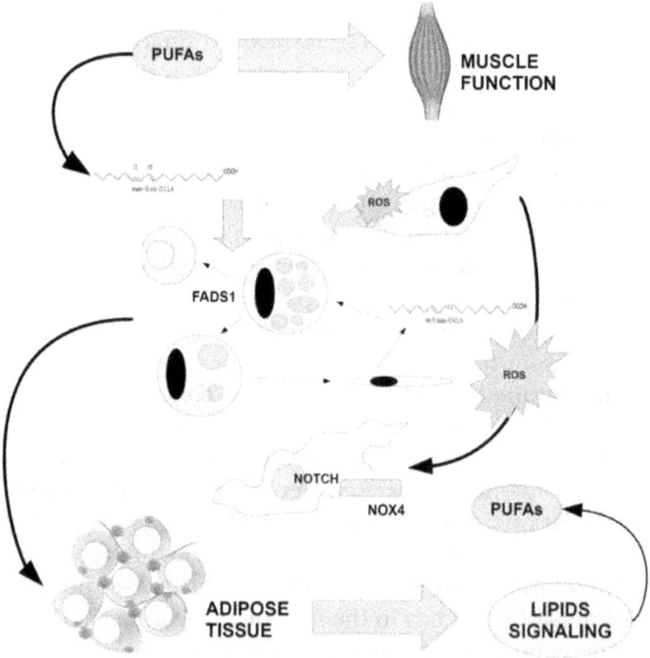

Figure 5.2 *Role of PUFAs in muscle development.*

5.3 Fatty Acids and Amino Acids

Amino acids are the bases for the synthesis of protein in the physiological processes and the majority of amino acids like glutamate, glutamine, and aspartate act as metabolic fuels in various organs of the fishes. Alkaline phosphatase too has been used as a potential biomarker in the fishes for the conversion of lipids. A biochemical fingerprint has been demonstrated in zooplankton species (Guisande et al., 2002; Boëchat & Adrian, 2005).

Diatom cultures are widely utilized for feeding shrimp larvae, zooplankton, juvenile oysters, and act as lipid sources, biofuel precursors, optical sensors in nanotechnology. The present work is a review of the different parameters such as pH, silica, temperature, and amino acids which encompass their effect on diatom growth. Variable temperature influenced the diatom growth rate, cell size, biochemical composition, and nutrient requirement. Alkaline pH results in an increase in triglyceride accumulation and acidic pH will result in the alteration of the nutrient uptake of diatoms. Thus, the growth study by altering different physical and chemical characteristics will provide a means to analyze and isolate biofuels, secondary metabolites, vitamins, carotenoids, and fatty acids from the diatoms. Calanus oil, a natural product that is extracted from marine zooplankton *Calanus finmarchicus*, is a novel type of bioactive n-3 PUFA lipid, containing substantial levels of wax esters (up to 80–90% of total lipids). Calanus oil is a highly bio-available source of eicosapentaenoic acid (EPA) and docosahexaenoic acid (DHA), fatty acids for human consumption with a beneficial effect on inflammation and obesity-related metabolic disturbances as shown in mice.

It's important to know how planktonic distributions, export fluxes, and biogeochemical cycling are affected by seasons around the Arabian Sea area. This creates a direct impact on the distribution of plankton which subsequently results in the nutrient change and therefore, the fishing community. Microzooplankton and mesozooplankton grazing are both very important in the high-nutrient coastal areas.

Calanus oil (COil) is a natural product extracted from marine zooplankton *Calanus finmarchicus* found in the North Atlantic Ocean. This oil is rich in wax esters of polyunsaturated fatty acids (PUFAs) and has been projected as the best alternative to fish oil because fish oil production cannot keep pace with the demands from the growing markets. The COil is the only commercially available marine source of wax esters, whereas classic ω-3 PUFAs come from triglycerides, ethyl esters, and phospholipids. There is a rise in the use of PUFA-rich oil in the aquaculture industry within the recent decade. A simultaneous rise in the demand for PUFAs is also observed in the health care industry for the use in prevention of lifestyle disorders—viz., obesity, diabetes mellitus, atherosclerosis, brain functioning, chronic low-grade inflammation, and cardiovascular disorders (CVDs).

There have been studies related to the health benefits of seafood and marine lipids. Marine animals are a great source for lipid-rich oil, with marine fishes being a common source for nutrient-rich oil. The lipid-rich oil from fishes is commonly used in aquaculture. Fish oil in aquaculture as a feed source is very common as it contains a very high quantity of health-promoting ω-3 long-chain polyunsaturated fatty acids (φ-3 LC-PUFAs) (Turchini et al., 2010). Scientists investigate currently, as an alternative to fish oils, oil-seed plants, and zooplankton to produce n-3 LCPUFAs. For the development of humans as well as for cancer prevention, the ratio of ω-6/ω-3 essential fatty acids in natural oil-rich products is very important.

Research has shown that oils rich in ω-3 PUFAs, such as eicosapentaenoic (EPA) and docosahexaenoic (DHA) acids, reduce the risk of cardiovascular disorders. Beneficial effects of these oils have also been seen in weight management, treatment of dyslipidemia, hypertension, diabetes mellitus, chronic low-grade inflammation, and atherosclerosis (Garcia-Esquinas et al., 2019; Turchini et al., 2010).

Various health-promoting benefits of the COil extracted from *C. finmarchicus* have been studied. *C. finmarchicus*, as a source of COil, is a widespread small marine crustacean constituting the major fraction of the zooplankton biomass presented in the North Atlantic Ocean. Triglycerides fulfill the short-term energy needs of these zooplanktons; wax esters serve as a long-term energy source for these small organisms. Wax esters are significant principles of the marine food chain since zooplankton is considered an important element of marine food webs. Several methods can be applied to obtain oil from marine biomass (e.g. wet or dry rendering, hydrolysis, silage production, supercritical fluid extraction, and solvent extraction) (Table 5.1, Pedersen et al., 2014).

The major active principles in the oil extracted from *C. finmarchicus* are exemplified by monoesters of LC-PUFA and fatty alcohols, namely wax esters. The Calanus oil is also rich in deep red carotenoid astaxanthin (ASX), which exhibits antioxidant effects (Figure 5.3, Gasmi et al., 2020). The fatty acid residues of its wax esters contain high amounts of stearidonic acid (SDA), EPA, and DHA, and also monounsaturated fatty acids (Walker et al., 2013). The structure of the characteristic wax ester of *C. finmarchicus* lipids, comprising fatty alcohol docosanol (22:1n-11) and PUFA. Substantial evidence shows that consumption and increased blood levels of the very LC ω-3 PUFAs (EPA, DHA, and α-linolenic acid) are associated with health benefits (Figure 5.4, Walker et al., 2013).

Table 5.1 Lipid Class Composition of Late Copepodite Stages and Adult *C. finmarchicus* Sampled in Different Periods and Presented as % Total Lipids % of Total Lipid

Lipid Class	June[1]	October[1]	March[1]	March[2]
Triacylglycerols	8.9	1.3	nd	3.1
Sterols	1.2	2.6	3.2	1.4
Free fatty acids	0.2	nd	1.7	Nd
Wax esters	85.4	88.1	84.9	73.8
Phospholipids	4.2	7.3	10.3	21

nd = not detected.
Source: Falk-Petersen et al. (1987); Pedersen et al. (2014).

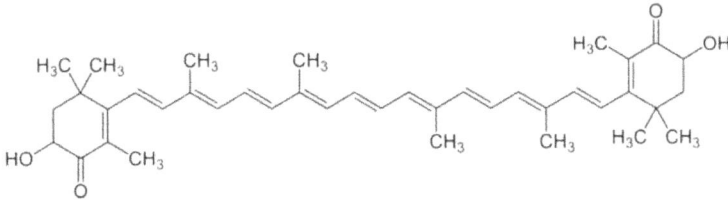

Figure 5.3 *The structural formula of carotenoid astaxanthin.*

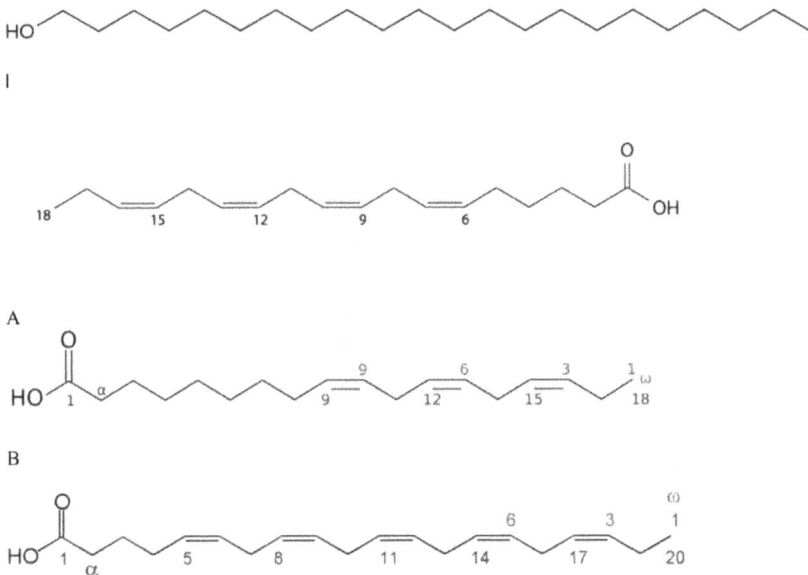

Figure 5.4 *Structural formulas of docosenol (1) and the PUFA (a stearidonic acid; b eicosapentaenoic acid; c docosahexaenoic acid) as a wax ester component of COil.*

Source: Gasmi et al. (2020)

Triacylglycerols containing mainly the ω-3 fatty acids might reach up to 8.9% and phospholipids up to 10.3% of total lipids in COil (Pedersen et al., 2014). Triglyceride esters of EPA and DHA are common in animal dietary products, fish oil, and COil (Horrocks & Yeo, 1999; Nakamura et al., 2014). DHA has an important impact on the growth and development of the infant's brain. The intake of DHA enhances learning skills, whereas its deficiency may lead to learning disability (Gasmi et al., 2020). DHA and EPA have a positive effect on diseases such as hypertension, arthritis, atherosclerosis, and depression. Also, DHA is needed in adults to maintain a normal function of the brain (Cook et al., 2016). The intense red color of COil is due to high carotenoid astaxanthin content (Pedersen, 2016). In zooplankton, ASX is the most commonly occurring carotenoid and may contribute to as much as 85–90% of the total pigment

(Ambati et al., 2014; Davinelli et al., 2018). Copepods use β-carotene, obtained from phytoplankton, as a precursor for ASX synthesis (Andersson et al., 2003). The specific structure of the ASX molecule provides its ability to be esterified, higher antioxidative capacity, and a more polar configuration than other carotenoids (Guerin et al., 2003). Thus, COil is the only commercially available marine source of wax esters, whereas classic ω-3s come as triglycerides, ethyl esters, and phospholipids. The wax esters in COil are slowly but completely digested and absorbed in the distal intestine. Digestion of wax esters releases the unsaturated fatty acids in the colon and activates the GPR120 receptors in immune cells that secrete hormones, which control the sugar and fat metabolism (O'Connell et al., 2017).

Of importance to be mentioned here is that the oil content of *C. finmarchicus* is regulated by the growth stage, location, and seasonal variations (Pedersen et al., 2014). The wax ester content is highest (88%) in *C. finmarchicus* during late autumn when the organism undergoes the feeding cycle and lowest (85% of the total lipids) during the winter season when the lipids are used as an energy source in gonad production (Kvile et al., 2016). Recent studies have reported that stage IV and stage V are the best developmental stages to extract the oil from *C. finmarchicus* as, during these stages, the lipids constitute as much as 60% of the total dry bodyweight of the organism (Bailey et al., 2012; Pedersen et al., 2014). The composition of wax esters vary between Calanus species (Graeve & Kattner, 1992; Pedersen et al., 2014). For example, lipids of *C. hyperboreus* are predominantly di-unsaturated long-chain wax esters while the lipids of *C. finmarchicus* are rich in monounsaturated short-chain wax esters (Graeve & Kattner, 1992).

Calanus oil recovers metabolic flexibility and rescues postischemic cardiac function accompanied by a significantly better recovery of cardiac performance following ischemia-reperfusion. Dietary supplementation with a small amount of oil from the marine crustacean *Calanus finmarchicus* reduces both intra-abdominal and hepatic fat deposition while at the same time exerting a strong anti-inflammatory action in adipose tissue during high-fat feeding of the diet.

If proteins are the main component of zooplankton biomass in all oceanographic regions, from the tropics to polar areas, they are closely followed by lipids (e.g. Percy & Fife, 1981; Donnelly et al., 1994; Kumar et al., 2013; Yun et al., 2015). Lipids in zooplankton organisms are very variable geographically, showing a latitudinal pattern with high 305 percentages in polar areas and low percentages in warm tropical waters, but also seasonal features, with higher percentages in summer than in winter (Falk-Petersen et al., 1999; Mayzaud et al., 2011; Kumar et al., 2013). Foraminifera of different species or within the same species have the potential to balance the ratio between their protein and lipid concentrations (known to be a major component of zooplankton Corg) in order to adapt to their environmental conditions (e.g. temperature change).

Cowie and Hedges (1996) studied the digestion in herbivorous zooplankton which feed upon the phytoplankton. They showed the quantitative estimation of the fate of aldoses and amino acids. Organic carbon and nitrogen are derived from aldoses, and particularly amino acids as important sources to the copepods.

Animals adapt to their habitat environments to survive when faced with specific environmental pressures. Importantly, changes in amino acid composition (AAC). The animals attempt to adapt to their habitats for survival in response to specific environmental pressures. AAC in the pelagic fishes is species-specific (Riveiro et al., 2011; Falco et al., 2016). Moreover, aquatic animals of species respond consistently to climate changes (i.e. a specie is adapted equally well as other species to different environmental conditions) (Garland & Kelly, 2006). Nonetheless, intra-species variations as a result of environmental changes were observed in the present study in agreement with other studies (Guisande et al., 1998; Riveiro et al., 2000, 2003). Among the considered environmental factors that are able to induce physiological stress, no significant differences were observed in the salinity of the water, while significant differences were recorded for the temperature and CHLsat. Lipid or carbohydrate content in the larvae did not differ between the individuals collected from the AB and MB areas. Early life stages of marine fish are particularly sensitive to environmental stressors, due to either the lack or low functional capacity of some organs and to the high metabolic rates needed for ATP production as well as growth and development (Pimentel et al., 2015). The higher activity of alkaline phosphatase in the larvae are due to more exposure to a higher level of stress.

Ederington et al. (1995) is a pioneer in studying the fatty acids and sterol transfer in the trophic levels between bacteria to copepods. It has been proven by Boëchat and Adrian (2005) that the enrichment of food quality was achieved by algae-derived biochemicals like PUFAs, sterols, and the amino acids. The nutrient enrichment in the zooplankton is more affected by the algal feed rather than any other kind of food they consume. The synthesis or conversion is an important aspect when it comes to the food chain. The interactions between the nutritional biochemicals and non-food biochemicals which involve toxins as well as aldehydes play a crucial role in the food chain (Adolf et al., 2006; Ianora et al., 2004). Free amino acids are used by copepods to balance the osmoregulation stresses generated by salinity changes (McAllen, 2003). The pool of free amino acids has been proportionate with the salinity.

The protozoans are qualitatively important for nourishment of the zooplankton for their low C:N ratios as compared to phytoplankton (Stoecker & Capuzzo, 1990). Fatty acids are the principal form of stored energy in many organisms. PUFAs seem to play an important role in cell membrane activity as precursors of prostaglandins, the hormones regulating ionic fluxes, oocyte maturation and egg production in invertebrates (Harrison, 1990). Taxonomic differences in fatty acid composition have been well established in the case of autotrophic phytoplankton. Heterotrophic dinoflagellates were found to be enhanced with fatty acid 22:6ω3 (Volkman, 1989). Zooplankton can be exploited as nutritional live feed in aquaculture for their absorbance or required carbohydrates and lipids. Prado-Cabrero and Nolan (2021) have also reported the importance of zooplankton-containing feed in aquaculture as a substitute to fishmeal and fish oil. This is because the zooplankton-containing food maintains the growth as compared to artificially made meal. Panigrahi and Jithendran (2021) have emphasized the use of brackishwater systems for nutritional value as a component of seafood species like fish, shrimp, crabs, etc. The Central Institute of Brackishwater Aquaculture (CIBA) also emphasizes the use of such feed for the development of shrimp and other organisms. These not only provide the essential amino acids, carotenoids, pigments, omega-3 fatty acids, vitamins, polysaccharides but also enhance the nutrient values of organisms belonging to subsequent trophic levels.

The report by Wacker and Von Elert (2001) was important because they could show a single biochemical substance (EPA or αLA) in the development of Daphnia (i.e. trophic transfer). Experiments with artificially supplemented single fatty acids report that EPA is a limiting factor for the growth of Daphnia. Cholesterol is vital for arthropods. Biochemical resources limit trophic relationships via zooplankton production and pelagic community structure. There are differences which occur between cladocerans and copepods with DHA in observations. Copepods and Daphnids have inversely proportional amounts of EPA and DHA and both fatty acids are important nutritional food constituents for calanoid copepods Heterotrophic plankton are vital food components of zooplankton for the availability of bacterial and biochemical composition of carbon. In essence, when describing food web interactions it is important to know whether synthesis or conversion takes place in a certain food web compartments, and if so to what degree. If the conversion effect is strong enough to alter the profile of the biochemical under consideration, its transfer would be affected. However, not only the biochemical make-up of the food and the metabolic capabilities and requirements of the consumers are important for the transfer of biochemical resources and energy within the pelagial system.

Free amino acids are used by copepods to balance the osmoregulation stresses generated by salinity changes (Farmer & Reeve, 1978; McAllen, 2003). The pool of FAA tend to increase with the salinity and it is often accepted that mostly non-essential amino acids (NEAA) are involved. There were no salinity changes involved in the present experiments (30 g L 1) and so differences in FAA contents between the organisms for this reason were not expected. The proportion of EAA has been observed to be higher in a rotifer than in two copepods when raised on the same food conditions. Certain types of fatty acids are considered essential (EFA) as they cannot be easily synthesized by the organism and must be obtained in sufficient quantity from the food to maintain growth and survival (Sargent & FalkPetersen, 1988; Harrison, 1990). EFA can be obtained directly from phytoplankton, but as our analysis and others show (e.g. Volkman, 1989), the different protists vary greatly in their EFA content. PUFAs seem to play an important role in cell membrane activity as precursors of prostaglandins, the hormones regulating ionic fluxes, oocyte maturation, and egg production in invertebrates (Harrison, 1990). Taxonomic differences in fatty acid composition have been well established in the case of autotrophic phytoplankton (e.g. Volkman, 1989). Heterotrophic dinoflagellates are particularly rich in the fatty acid 22:6ω3 and that their fatty acid composition does not differ significantly from that of autotrophic dinoflagellates (Volkman, 1989). The inconclusive result of the residual analysis could be either the consequence of a species-specific difference in fatty acid requirement, or of a need for other essential nutritional components not measured here (e.g. sterols, amino acids, and proteins), which appear to be implicated in regulating copepod egg production (for sterols, e.g. Ederington et al., 1995; for amino acids, e.g. Guisande et al., 2000). Reproduction success of copepod populations depends not only on fecundity rates but also on the viability of eggs. Similarly, the study of Ederington et al. (1995) has shown high hatching success for *A. tonsa* fed either diatoms.

African catfish (*Clarias gariepinus*) is a slow-moving omnivorous predatory fish, which feeds on a variety of food items from microscopic zooplankton to fish half its length or 10% of its own body weight. African catfish is of very high demand in the middle belt and north east of Nigeria on account of their tasty flesh. The cyclic order of the mercury contamination chain starts from its emission in industries. This is followed by contamination in the atmosphere, soil, water, phytoplankton, zooplankton, fish, and humans. There are two important sources of mercury: anthropogenic and natural sources. Estimation of biochemical composition of zooplankton is important in understanding their metabolism, nutritive value, and energy transfer which are relevant to the marine ecosystem. Zooplankton biomass and their biochemical composition have been estimated where protein is found to be a major fraction of the organic constituents. Zooplankton can be utilized as nutritional live feed for the cultivable species of fish and prawn in aquaculture farms. The variations in biochemical composition of zooplankton are influenced by species composition and feeding activities of zooplankton, which is in accordance with the previous studies.

Prado-Cabrero and Nolan (2021) indicate that aquaculture is looking for substitutes for fishmeal and fish oil to maintain its continued growth. Zooplankton is the most nutritious option, but its controlled mass production has not yet been achieved. Because of fast growth over the last 20 years, fed aquaculture has recently outpaced non-fed aquaculture production of aquatic animals. Zooplankton is the most nutritious feed for fish, but nowadays, its use to replace fish meal and fish oil is unthinkable. Artemia eggs obtained from natural ponds, copepod eggs produced by various companies, and rotifers produced on-site are used in hatcheries; mass and controlled production of zooplankton has not yet been achieved.

References

Adolf, J. E., Yeager, C. L., Miller, W. D., Mallonee, M. E., & Harding Jr, L. W. (2006). Environmental forcing of phytoplankton floral composition, biomass, and primary productivity in Chesapeake Bay, USA. *Estuarine, Coastal and Shelf Science, 67*(1–2), 108–122.

Ambati, R. R., Phang, S. M., Ravi, S., & Aswathanarayana, R. G. (2014). Astaxanthin: sources, extraction, stability, biological activities and its commercial applications—a review. *Marine drugs, 12*(1), 128–152.

Andersson, M., Van Nieuwerburgh, L., & Snoeijs, P. (2003). Pigment transfer from phytoplankton to zooplankton with emphasis on astaxanthin production in the Baltic Sea food web. *The Marine Ecology Progress Series, 254*, 213–224.

Bailey, C., McMeans, B. C., Arts, M. T., Rush, S. A., & Fisk, A. T. (2012). Seasonal patterns in fatty acids of Calanus hyperboreus (Copepoda, Calanoida) from Cumberland Sound, Baffin Island, Nunavut. *Mar Biol, 159*(5), 1095–1105.

Boëchat, I. G., & Adrian, R. (2005). Biochemical composition of algivorous freshwater ciliates: you are not what you eat. *FEMS Microbiology Ecology, 53*(3), 393–400.

Cook, C. M., Larsen, T. S., Derrig, L. D., Kelly, K. M., & Tande, K. S. (2016). Wax ester rich oil from the marine crustacean, Calanus finmarchicus, is a bioavailable source of EPA and DHA for human consumption. *Lipids, 51*(10), 1137–1144.

Cowie, G. L., & Hedges, J. I. (1984). Carbohydrate sources in a coastal marine environment. *Geochimica et Cosmochimica Acta, 48*, 2075–2087.

Cowie, G. L., & Hedges, J. I. (1996). Digestion and alteration of the biochemical constituents of a diatom (*Thalassiosira weisflogii*) ingested by an herbivorous zooplankton (*Calanus pacificus*). *Limnology and Oceanography, 41*(4), 581–594.

Davinelli, S., Nielsen, M. E., & Scapagnini, G. (2018). Astaxanthin in skin health, repair, and disease: a comprehensive review. *Nutrients, 10*(4), 522.

Donnelly, J., Torres, J. J., Hopkins, T. L., & Lancraft, T. M. (1994). Chemical composition of Antarctic zooplankton during austral fall and winter. *Polar Biology, 14*(3), 171–183.

Ederington, M. C., McManus, G. B., & Harvey, H. R. (1995). Trophic transfer of fatty acids, sterols, and a triterpenoid alcohol between bacteria, a ciliate, and the copepod Acartia tonsa. *Limnology and Oceanography, 40*(5), 860–867.

Falco, F., Barra, M., Cammarata, M., Cuttitta, A., Jia, S., Bonanno, A., Mazzola, S, & Wu, G. (2016). Amino acid composition in eyes from zebrafish (*Danio rerio*) and sardine (*Sardina pilchardus*) at the larval stage. *SpringerPlus, 5*, 519. https://doi.org/10.1186/s40064- 016-2137-1

Falk-Petersen, S., Sargent, J. R., Loenne, O. J., & Timofeev, S. (1999). Functional biodiversity of lipids in Antarctic zooplankton: *Calanoides acutus, Calanus propinquus, Thysanoessa macrura and Euphausia crystallorophias. Polar Biology, 21*, 37–47.

Falk-Petersen, S., Sargent, J. R., & Tande, K. S. (1987). Lipid composition of zooplankton in relation to the sub-arctic food web. *Polar Biology, 8*(2), 115–120.

Farmer, L., & Reeve, M. R. (1978). Role of the free amino acid pool of the copepod Acartia tonsa in adjustment to salinity change. *Marine Biology, 48*, 311–316.

Garland, T., & Kelly, S. A. (2006). Phenotypic plasticity and experimental evolution. *Journal of Evolutionary Biology, 209*, 2344–2361.

Gasmi, A., Mujawdiya, P. K., Shanaida, M., Ongenae, A., Lysiuk, R., Doşa, M. D., . . . & Björklund, G. (2020). Calanus oil in the treatment of obesity-related low-grade inflammation, insulin resistance, and atherosclerosis. *Applied Microbiology and Biotechnology, 104*(3), 967–979.

Graeve, M., & Kattner, G. (1992). Species-specific differences in intact wax esters of Calanus hyperboreus and C. finmarchicus from Fram Strait—Greenland Sea. *Marine Chemistry, 39*(4), 269–281.

Guisande, C., Maneiro, I., Riveiro, I., Barreiro, A., & Pazos, Y. (2002). Estimation of copepod trophic-niche in the field using amino acids and marker pigments. *Marine Ecology Progress Series, 239*, 147–156.

Guisande, C., Riveiro, I., Sola, A., & Valdes, L. (1998). Effect of biotic and abiotic factors in the biochemical composition of wild eggs and larvae of several fish species. *Marine Ecology Progress Series, 163,* 53–61.

Guerin, M., Huntley, M. E., & Olaizola, M. (2003). Haematococcus astaxanthin: applications for human health and nutrition. *Trends Biotechnology, 21*(5), 210–216.

Harrison, K. E. (1990). The role of nutrition in maturation, reproduction and embryonic development of decapod crustaceans: a review. *Journal of Shellfish Research, 9,* 1–28.

Harvey, H. R., Eglinton, G., O'Hara, S. C., & Corner, E. D. (1987). Biotransformation and assimilation of dietary lipids by Calanus feeding on a dinoflagellate. *Geochimica et Cosmochimica Acta, 51*(11), 3031–3040.

Horrocks, L. A., & Yeo, Y. K. (1999). Health benefits of docosahexaenoic acid (DHA). *Pharmacological Research, 40*(3), 211–225.

Ianora, A., Miralto, A., Poulet, S. A., Carotenuto, Y., Buttino, I., Romano, G., . . . & Smetacek, V. (2004). Aldehyde suppression of copepod recruitment in blooms of a ubiquitous planktonic diatom. *Nature, 429*(6990), 403–407.

Josué, I. I., Cardoso, S. J., Miranda, M., Mucci, M., Ger, K. A., Roland, F., & Marinho, M. M. (2019). Cyanobacteria dominance drives zooplankton functional dispersion. *Hydrobiologia, 831*(1), 149–161.

Kumar, M. A., Padmavati, G., & Anandavelu, I. (2013). Biochemical composition and calorific value of zooplankton from the coastal waters of South Andaman. *Proceedings of the International Academy of Ecology and Environmental Sciences, 3*(3), 278–287.

Kvile, K. O., Langangen, O., Prokopchuk, I., Stenseth, N. C., & Stige, L. C. (2016). Disentangling the mechanisms behind climate effects on zooplankton. *Proceedings of the National Academy of Sciences of the United States of America, 113*(7), 1841–1846.

Leibold, M. A. (1995). The niche concept revisited: mechanistic models and community context. *Ecology, 76*(5), 1371–1382.

Mayzaud, P., Lacombre, S., & Boutoute, M. (2011). Seasonal and growth stage changes in lipid and fatty acid composition in the multigeneration copepod Drepanus pectinatus from Iles Kerguelen. *Antarctic Science, 23,* 3–17.

McAllen, R. (2003). Variation in the free amino acid concentrations of the supralittoral rock-pool copepod crustacean Tigriopus brevicornis during osmotic stress. *Journal of the Marine Biological Association of the United Kingdom, 83*(5), 921–922.

Nakamura, M. T., Yudell, B. E., & Loor, J. J. (2014). Regulation of energy metabolism by long-chain fatty acids. *Progress in Lipid Research, 53,* 124–144.

O'Connell, T. D., Block, R. C., Huang, S. P., & Shearer, G. C. (2017) Omega-3-polyunsaturated fatty acids for heart failure: effects of dose on efficacy and novel signaling through free fatty acid receptor 4. *Journal of Molecular and Cellular Cardiology, 103,* 74–92. https://doi.org/10.1016/j.yjmcc.2016.12.003

Panigrahi, A., & Jithendran K. P. (2021). Innovative culture practices in brackishwater aquaculture with ecosystem approach. *Training Manual on Biofloc Technology for Nursery and Growout Aquaculture.*

Pedersen, A. M. (2016). Calanus® oil. Utilization, composition and digestion. Doctoral thesis. Arctic University of Norway, Tromsø.

Pedersen, A. M., Vang, B., & Olsen, R. L. (2014). Oil from Calanus finmarchicus—composition and possible use: a review. *Journal of Aquatic Food Product Technology, 23*(6), 633–646.

Percy, J. A., & Fife, F. J. (1981). The biochemical composition and energy content of Arctic marine andomizednkton. *Arctic, 34,* 307–313.

Pimentel, M. S., Faleiro, F., Diniz, M., Machado, J., Pousão-Ferreira, P., Peck, M. A., . . . & Rosa, R. (2015). Oxidative stress and digestive enzyme activity of flatfish larvae in a changing ocean. *PloS One, 10*(7), e0134082.

Prado-Cabrero, A., & Nolan, J. M. (2021). Omega-3 nutraceuticals, climate change and threats to the environment: the cases of Antarctic krill and Calanus finmarchicus. *Ambio, 50*(6), 1184–1199.

Prahl, F. G., Muehlhausen, L. A., & Zahnle, D. L. (1988). Further evaluation of long-chain alkenones as indicators of paleoceanographic conditions. *Geochimica et Cosmochimica Acta, 52*(9), 2303–2310.

Ramlee, A., Chembaruthy, M., Gunaseelan, H., Yatim, S. R. M., Taufek, H., & Rasdi, N. W. (2021, November). Enhancement of nutritional value on zooplankton by alteration of algal media composition: a review. In *IOP conference series: earth and environmental science* (Vol. 869, No. 1, p. 012006).

Riveiro, I., Guisande, C., Franco, C., Lago De Lanzós, A., Maneiro I., & Vergara, A. R. (2003). Egg and larval amino acid composition as indicators of niche resource partitioning in pelagic fish species. *Marine Ecology Progress Series, 260*, 255–262.

Riveiro, I., Guisande, C., Iglesias, P., Basilone, G., Cuttitta, A., Giráldez, A., Patti, B., Mazzola, S., Bonanno, A., Vergara, A. R., & Maneiro, I. (2011). Identification of subpopulations in pelagic marine fish species using amino acid composition. *Hydrobiologia, 670*, 189–199.

Riveiro, I., Guisande, C., Lloves, M., Maneiro, I., & Cabanas, J. M. (2000). Importance of parental effects on larval survival in Sardina pilchardus. *Marine Ecology Progress Series, 205*, 249–258.

Sargent, J. R., & Falk-Petersen, S. (1988). The lipid biochemistry of calanoid copepods. *Hydrobiologia, 167/168*, 101–114.

Sherr, E. B., & Sherr, B. F. (2002). Significance of predation by protists in aquatic microbial food webs. *Antonie Van Leeuwenhoek, 81*(1), 293–308.

Stoecker, D. K., & Capuzzo, J. M. (1990). Predation on protozoa: its importance to zooplankton. *Journal of Plankton Research, 12*(5), 891–908.

Tou, J. C., Jaczynski, J., & Chen, Y. C. (2007). Krill for human consumption: nutritional value and potential health benefits. *Nutrition Reviews, 65*(2), 63–77.

Turchini, G. M., Ng, W. K., & Tocher, D. R. (2010). *Fish oil replacement and alternative lipid sources in aquaculture feeds*. CRC Press.

Volkman, J. K. (1989). Fatty acids of microalgae used as feedstocks in aquaculture. In: Cambie, R. C. & E. Horwood (eds), *Fats for the future*. Chichester: 263–283.

Volkman, J. K., Jeffrey, S. W., Nichols, P. D., Rogers, G. I., & Garland, C. D. (1989). Fatty acid and lipid composition of 10 species of microalgae used in mariculture. *Journal of Experimental Marine Biology and Ecology, 128*(3), 219–240.

Wacker, A., & Von Elert, E. (2001). Polyunsaturated fatty acids: evidence for non-substitutable biochemical resources in Daphnia galeata. *Ecology, 82*, 2507–2520.

Wakeham, S. G., & Lee, C. (1993). Production, transport, and alteration of particulate organic matter in the marine water column. In *Organic geochemistry* (pp. 145–169). Springer, Boston, MA.

Walker, C. G., Jebb, S. A., & Calder, P. C. (2013). Stearidonic acid as a supplemental source of ω-3 polyunsaturated fatty acids to enhance status for improved human health. *Nutrition, 29*(2), 363–369.

Yun, M. S., Lee, D. B., Kim, B. K., Kang, J. J., Lee, J. H., Yang, E. J., Park, W. G., Chung, K. H., & Lee, S. H. (2015). Comparison of phytoplankton macromolecular compositions and zooplankton proximate compositions in the northern Chukchi Sea. *Deep Sea Research Part II, 120*, 82–90.

Yuslan, A., Nasir, N., Suhaimi, H., Arshad, A., & Rasdi, N. W. (2021, November). The effect of enriched cyclopoid copepods on the coloration and feeding rate of Betta splendens. In *IOP conference series: earth and environmental science* (Vol. 869, No. 1, p. 012007). IOP Publishing.

Nutraceuticals from Mollusks

Gandhi Rádis-Baptista

Contents

6.1 Introduction

Mollusks consist of invertebrate animals with an evolutionary history that dates back to 550 million years ago. The phylum Mollusca is the second largest in the animal kingdom with over 80,000 living species, predominantly marine, which are grouped in seven classes based on the morphology of their foot and presence or absence and type of shell (Wallace and Taylor 2003). The most known classes of mollusks comprise the Bivalvia or Pelecypoda (cockles, mussels, oysters, scallops and clams), Cephalopoda (squids, octopuses and cuttlefishes) and Gastropoda (snail and snail-like mollusks). Notably, the gastropods and bivalves are the first and the second most widely diverse and numerically largest species that coincidently include mollusk species that comprise

DOI: 10.1201/9781003128175-6

seafood of nutritional and economic importance. The nutritional value and health benefits of species of edible and commercially valuable mollusks have been noticed since ancient times; seen such as a food and nutrient resource they have been tamed and used by humans for generations in different cultures (De Zoysa 2012). Concerning the nutritional aspects of mollusks, the Food and Agricultural Organization of United Nations (FAO) compiled and organized nutrient data related to the energy, macronutrients, main minerals and vitamins, amino acids and fatty acids of raw and cooked mollusks through the FAO/INFOODS global food composition database for fish and shellfish Version 1.0 (uFiSh1.0) (FAO/INFOODS 2016). In Table 6.1, species of edible mollusks reported as seafood, and that appear in the uFiSh1.0, are listed.

The nutrient content and ingredient composition of mollusks as seafoods are variable and closely associated to the species, sex, age, seasonality variations, food availability and abiotic factors (e.g., salinity and water temperature), as well as related to the process of food preparation (FAO/INFOODS 2016). Mollusks as seafoods are rich in lipids, polysaccharides, polypeptides (proteins), vitamins and minerals, which provide energy, metabolic regulators and chemical building blocks for biosynthesis of biomolecules in the human body.

Table 6.1 Mollusk Species in the Food List of the Fish and Shellfish Database (uFiSh1.0) of the Food and Agricultural Organization of United Nations

Class	Family	Species	Common Name
Bivalvia	Ostreidae	*Crassostrea* spp	Cupped oyster
		Crassostrea gigas	Cupped cupped oyster
		Crassostrea rhizophorae	Mangrove cupped oyster
		Crassostrea virginica	American cupped oyster
		Ostrea spp	Flat oyster
		Ostrea edulis	European flat oyster
	Mytilidae	*Mytilus* spp	Mytilus mussel
		Mytilus edulis	Blue mussel
		Mytilus galloprovincialis	Mediterranean mussel
		Perna spp	Perna mussel
		Perna canaliculus	New Zealand mussel
		Perna viridis	Green mussel
	Pectinidae	*Diverse species of this family*	Scallops
		Pecten maximus	Great Atlantic scallop
	Veneridae	*Diverse species of this family*	Venus clams
		Chamelea gallina	Striped venus
Gastropoda	Haliotidae	*Haliotis* spp	Abalones
	Strombidae	*Diverse species of this family*	Conch shells
		Strombus galeatus	Giant Eastern Pacific conch
		Strombus gracilior	Eastern Pacific fighting conch

Table 6.1 Mollusk Species in the Food List of the Fish and Shellfish Database (uFiSh1.0) of the Food and Agricultural Organization of United Nations. Continued

Class	Family	Species	Common Name
Cephalopoda	Loliginidae	*Loligo spp*	Inshore squids
		Loligo vulgaris	European squid
	Octopodidae	*Octopus spp*	Octopuses
		Octopus vulgaris	Common octopus
	Ommastrephidae	*Diverse species of this family*	Ommastrephidae squids
	Sepiidae, Sepiolidae	*Diverse species of these families*	Cuttlefish, bobtail squids
	Sepiidae	*Sepia officianlis*	Common cuttlefish

Note: Edible mollusks listed in the FAO/INFOODS global food composition database for fish and shellfish Version 1.0 (uFiSh1.0) (FAO/INFOODS 2016), in which nutrient data related to the energy, macronutrients, main minerals and vitamins, amino acids and fatty acids composition of raw and cooked mollusks are compiled.

Notably, besides comprising significant food resources with high nutrient value, mollusks contain many unique bioactive ingredients with health-promoting properties. These bioactive ingredients impart to molluscan-derived products the features of functional foods. Mollusks as functional foods are naturally augmented with polyunsaturated fatty acids (PUFA), sterols and their precursors (e.g., isoprenoids), fat-soluble vitamins (vitamins A, E, D and K), polysaccharides and complex glycans (e.g., glycosaminoglycans), metalloproteins (e.g., hemocyanin) and bioactive peptides (e.g., antimicrobial peptides in the hemolymph and protein fragments in protein hydrolysates), all of which have beneficial impacts on human health beyond their nutritive value.

Moreover, mollusks contain some special classes of biologically and pharmacologically active compounds that possess functional health-promoting and disease-preventing properties, as well as the capability of ameliorate pathophysiological conditions of certain disorders. These molluscan-derived active ingredients are qualified to be converted into nutraceuticals in a variety of pharmaceutical forms. Indeed, the health-promoting efficacy of marine mollusks has been reviewed, and experimental data reported, regarding the antimicrobial, antiviral, anti-inflammatory and immunomodulatory properties of extracts, biochemicals and natural products from diverse species of bivalves, gastropods and cephalopods (Khan and Liu 2019). The bioactive ingredients from mollusks have been prepared by a number of selected techniques that recover purified compounds (i.e., biomolecules and organics) and extracts with antioxidant, anticancer, anti-infectious and cardiovascular-protecting properties, to mention a few (Odeleye, White, and Lu 2019).

Mollusks, in different preparation forms, have been used in traditional medicine for a long time, mainly in Asian countries, and with the advent of aquaculture these marine organisms emerge as unparalleled resources of bioactive compounds and natural products for the cosmeceutical, functional food and nutraceutical industries. For instance, the anti-inflammatory, immunomodulatory and wound-healing properties of molluscan-derived products have been reviewed concerning their traditional and contemporary uses, with data from *in vitro*, *in vivo* and human clinical trials (Ahmad et al. 2018). The anticancer and cancer preventive properties of edible mollusks and other marine organisms have also been noticed with attention, particularly with regard to bioactive peptides which have antiproliferative and antimetastatic activities against certain types of cancer cells, including from breast, prostate, lung and colon cancers (Correia-da-Silva et al. 2017). Examples of crude powder extracts, hydrolysates, individual biomolecules and their chemical classes, and natural products derived from mollusks with biological and pharmacological properties are presented and discussed in this chapter.

6.2 Selected Compounds from Mollusks: Bioactivity and Medicinal Properties

The following sections highlight particular classes of biomolecules and substances with health-promoting and disease-preventing properties obtained from mollusks that comprise active ingredients for functional foods and nutraceuticals.

6.3 Protein Hydrolysates and Bioactive Peptides

Bioactive peptides from mollusks can be obtained either as individual, pure chemical entities, as components of tissue and protein hydrolysates or as chemical class-enriched extracts (e.g., lipid-rich and polysaccharide-rich) with a range of biological and pharmacological activities. For instance, protein hydrolysates of marine organisms and tissues, like fish skin, have significant functionalities for nutraceutical, pharmaceutical and cosmeceutical applications owing to their antioxidant, antimicrobial and anti-aging activities (Venkatesan et al. 2017). Likewise, an increasing number of hydrolysates from diverse mollusk species, mostly from bivalves, have shown health-promoting properties, with their specific biological activities and respective mechanisms of action on the health improvement investigated in molecular details *in vitro* and *in vivo*. Hydrolysates from molluscan tissues with antioxidant (He et al. 2019, Yang et al. 2019), anti-hypertensive (Yu, Zhang, Luo, et al. 2018), anti-diabetic and anti-hyperlipidemic (Ben Slama-Ben Salem et al. 2018), cytoprotective (Oh, Ahn, Nam, et al. 2019), metal-chelating (Wu et al. 2019) and osteogenic (Xu, Zhao, et al. 2019), among other activities have been documented. Additionally, individual peptides from molluscan hydrolysates have been purified with medicinal properties that include anti-coagulant (Cheng et al. 2021; Cheng et al. 2018), anti-hypertensive (Chatterjee et al. 2020), antimicrobial (Harnedy and FitzGerald 2012; Grienke, Silke, and Tasdemir 2014; Wang et al. 2017), antioxidant (Yang et al. 2019; He et al. 2019), antitumoral (Suarez-Jimenez, Burgos-Hernandez, and Ezquerra-Brauer 2012), aphrodisiac (Zhang et al. 2019), metal-binding (Li, Gong, et al. 2019) and immunomodulatory activity (Li, Ye, et al. 2019). In Table 6.2, some bioactive

Table 6.2 Examples of Bioactive Peptides from Mollusks, Their Respective Functionalities, and Sources

Biological Activity and Peptide Sequence	Species	Reference
Anticoagulant and Antithrombotic		
ELEDSLDSER	Blue mussel, *Mytilus edulis*	(Qiao et al. 2018)
DFEEIPEEYLQ	Pacific oyster, *C. giga*	(Cheng et al. 2018)
NAESLRK	Pacific oyster, *C. giga*	(Cheng et al. 2021)
Anti-Hypertensive		
EVMAGNLYPG	Blue mussel, *M. edulis*	(Wang et al. 2017)
WPMGF	Sea snail, *Crassispira sinensis*	(Yu, Zhang, Luo, et al. 2018)
YSQLENEFDR	Marine snail, *Neptunea arthritica*	(Zhang, Han, et al. 2018)
Antimicrobial		
GFGCPNDYCHRHCKSIPGRXGGYCGGXHRLRCTCYR	Blue mussel, *M. edulis*	(Charlet et al. 1996)
GCASRCKAKCAGRRCKGWASASFRGRCYCKCFRC	Blue mussel, *M. edulis*	(Charlet et al. 1996)
SRAGLQFPVGRIHRLLRKGNYA	Oyster, *C. madrasensis*	(Sathyan et al. 2012)
DTFDYKKFGYRYDSLELEGRSISRIDELIQQRQEKD RTFAGFLLKGFGTSAS	Abalone, *Haliotis tuberculata*	(Zhuang et al. 2015)
GFCNFMHLKPISRELRRELYGRTRRRRK	Cuttlefish, *Sepia offcinalis*	(Benoist et al. 2020)
Antioxidant		
LKQELEDLLEKQE	Pacific oyster, *C. giga*	(Qian et al. 2008)
YPPAK	Blue mussel, *M. edulis*	(Wang et al. 2013)
WCTSVS	Indian squid, *Loligo duvauceli*	(Sudhakar and Nazeer 2015)
LSDRLEETGGASS, KEGCREPETEKGHR and IVTNWDDMEK	Hard clam, *Meretrix meretrix*	(Jia et al. 2018)
DTETGVPT	Abalone, *Volutharpa ampullacea*	(He et al. 2019)
IVVPK	Pacific oyster, *C. giga*	(Bang, Jin, and Choung 2020)
DVEDLEAGLAK and EITSLAPSTM	Golden cuttlefish, *Serpia esculenta*	(Yu et al. 2020)
Anti-Inflammatory and Immunomodulatory		
SCASRCKSRCRARRCRYYVSVRYGGFCYCRC	Med. mussel, *M. galloprovincialis*	(Balseiro et al. 2011)
GVSLLQQFFL	Korean mussel, *M. coruscus*	(Kim et al. 2013)
RVAPEEHPVEGRYLV	Sea snail, *C. sinensis*	(Li, Ye, et al. 2019)
EGLLGDVF	Green Mussel *Perna viridis*	(Joshi and Nazeer 2020b)
YA	Pacific oyster, *C. giga*	(Siregar et al. 2020)
HKGQCC	Saltwater clam, *Meretrix meretrix*	(Joshi, Mohideen, and Nazeer 2021)

(Continued)

Table 6.2 Examples of Bioactive Peptides from Mollusks, Their Respective Functionalities, and Sources. Continued

Biological Activity and Peptide Sequence	Species	Reference
Antiproliferative (Antitumoral)		
AFNIHNRNLL	Mussel, *M. coruscus*	(Kim et al. 2012)
HFQIGQRCLC	Pacific oyster, *C. giga*	(Cheong et al. 2013)
LKEENRRRRD	Golden cuttlefish, *S. esculenta*	(Huang et al. 2017)
ILYMP	Sea snail, *C. sinensis*	(Yu, Zhang, Ye, et al. 2018)
KVEPQDPSEW	Abalone, *H. discus hannai*	(Gong et al. 2019)
Metal-Binding		
HLRQEEKEEVTVGSLK	Pacific oyster, *C. giga*	(Chen et al. 2013)
EVPPEEH	Pacific oyster, *C. giga*	(Li, Gong, et al. 2019)
Osteogenic		
YRGDVVPK	Pacific oyster, *C. giga*	(Chen et al. 2019)
AWLNH and PHDL	Ark shell, *S. subcrenata*	(Oh, Ahn, Hyung, et al. 2019)
IEELEEELEAER	Blue mussels, *M. edulis*	(Xu, Chen, et al. 2019)

peptides, their functions, amino acid sequences and mollusk species in which they are found are indicated.

6.4 Anti-Hypertensive Peptides

Anti-hypertensive peptides are derived from a variety of foods, but mainly from milk (casein and whey), and animal collagen (and gelatin) hydrolysates have attracted attention in the last decades because of their availability, cost-effective production and benefits to human health. These food-derived anti-hypertensive peptides are short sequences of amino acid residues that mainly inhibit the activity of the angiotensin converting enzyme (ACE) in both the renin-angiotensin and kallikrein-quinine systems, which control the arterial tonus and, therefore, blood pressure (Miralles, Amigo, and Recio 2018). Anti-hypertensive peptides have been found in a variety of mollusk species, in crude hydrolysates or in pure form, as aforementioned and exemplified herein. For instance, from the protein hydrolysate of the pearl oyster *Pinctada fucata martensii*, two novel ACE inhibitory peptides were purified and their anti-hypertensive activity compared *in vitro* with the effect *in vivo* of the total meat hydrolysate. This comparison revealed the efficacy of such peptides to ameliorate hypertension and confirmed this mollusk source and peptides serve as functional food ingredients (Liu et al. 2019).

6.5 Antimicrobial Peptides

Antimicrobial peptides (AMPs) from molluscan tissues and hydrolysates are either gene-encoded, full-sequences that are released from cellular stores or fragments encrypted in full-size polypeptide chains that are unleashed after chemical or enzymatic cleavage of proteins. In the first case, gene-encoded AMPs represent host defense peptides that are components of the innate immunity of multicellular organisms used to fight against invading microbes. A diversity of structures with a narrow or broad-spectrum of action are known that include α-helical peptides, defensins (α, β and γ), cathelicidins, linear and proline-rich peptides, among others, which are active against bacteria, fungi, protozoa and even viruses (Zasloff 2002). Essentially, AMPs disrupt the membrane of microbial cells with a rapid onset of action, causing the leakage of cytoplasmic content followed by microbial cell death. Aside the membrane-disrupting activity, some AMPs exert their effects intracellularly (Le, Fang, and Sekaran 2017). Diverse classes of AMPs have been disclosed from bivalves and gastropods. For instance, defensins, littoreins, mytilins, mytimicins and molluscidins have been characterized from several species of bivalves and gastropods (Li et al. 2011; Valtchev 2018). In cephalopods, three novel linear peptides, namely NF19, AV19 and GK28, were undisclosed from the hemocyte extract of cuttlefish *Sepia officinalis* (Benoist et al. 2020). In the second case (i.e., peptides that are unleashed by chemical or enzymatic hydrolysis of proteins), peptides derived from histones-H2A (e.g., abhisin and molluskin) and hemocyanin (e.g., haliotisin) have been characterized with broad-spectrum antimicrobial activity (De Zoysa et al. 2009; Sathyan et al. 2012; Zhuang et al. 2015).

6.6 Anti-Inflammatory Peptides

From the foot tissue of the saltwater clam *Meretrix meretrix*, which is found in abundance in the Indian coastal areas, the anti-inflammatory hexapeptide NPAQAC was purified from protein hydrolysates that control the levels of pro-inflammatory cytokines and nitric oxide production, and suppress cyclooxygenase-2 (COX-2) activation in LPS-stimulated murine macrophage (RAW264.7) cells (Joshi and Nazeer 2020a). Likewise, the hexapeptide HKGQCG was purified from *M. meretrix* protein hydrolysates that originated from two shorter peptides by means of gastrointestinal digestion *in vitro*. The most active, thermostable short peptide possesses an effective anti-inflammatory property, acting through the inhibition of the nitric oxide/nitric oxide synthases (NO/iNOS) and prostaglandin G2/cyclooxygenase 2 (PGE2/COX-2) pathways and down-regulation of pro-inflammatory cytokines expression in lipopolysaccharide (LPS)-induced adult zebrafish (Joshi, Mohideen, and Nazeer 2021). An anti-inflammatory decapeptide (EGLLGDVF) was also purified from the protein hydrolysates of the foot tissue of the green mussel *Perna viridis* that restrained the production of pro-inflammatory cytokines, inhibited the

NO and COX-2 activation, and down-regulated the iNOS and COX-2 protein expression in LPS-stimulated RAW264.7 (Joshi and Nazeer 2020b). As another example, the dipeptide YA purified from the Pacific oyster *C. giga* suppresses cytochrome P450E1 (CYP2E1) enzyme, reactive oxygen species (ROS) production, and apoptotic and inflammatory mediators in liver tissues, in *in vivo* experiments. Interestingly, both the YA dipeptide and protein hydrolysate performs similarly when pre-administrated in mouse models of alcohol liver damage (Siregar et al. 2020).

6.7 Antioxidant Peptides

In hydrolysate prepared from the tissue of the Pacific oyster *Crassostrea gigas*, an abundant food resource in Asia and Europe, a range of peptides were identified that inhibit wrinkle formation, skin thickening and collagen degradation by means of downregulating the matrix metalloprotease expression via the regulation of the MAPK pathway in UVB-irradiated hairless mice (Han et al. 2019). From the protein hydrolysate of this oyster, an antioxidant pentapeptide (IVVPK) was identified that was shown to possess an anti-wrinkle effect in experimental animals (Bang, Jin, and Choung 2020) Antioxidant peptides with variable numbers of amino acid residues have been also isolated from the tissue and hydrolysates of other species of mollusks. For instance, from edible marine invertebrates that include oysters, mussels, clams, scallops and squids, an increasing number of antioxidant peptide sequences are being discovered through specific enzymatic digestions of molluscan extracts and by means of assay-guided purification and identification (Chai et al. 2017). At the molecular level, these peptides possess free-radical and alkyl-radical scavenger properties and protection against oxidative-induced DNA damage.

6.8 Antiproliferative Peptides

Protein hydrolysates and bioactive peptides with antitumor properties have been disclosed from a variety of food resources (Ifeanyi and Rotimi 2019). The anticancer activity and efficacy of food-derived protein hydrolysates and individual peptides have been experimentally tested *in vitro*, in cell culture, and *in vivo*, in animal models of tumor (Chalamaiah, Yu, and Wu 2018). Whatever the food origin, anticancer peptides differ in amino acid composition, sequence and length. An example of the shortest antiproliferative peptide characterized from mollusk is a pentapeptide ILYMP obtained from the protein hydrolysate of the sea snail *Cyclina sinensis*. This antiproliferative pentapeptide induced apoptosis in prostate cancer (DU-145) cells via up-regulation of Bcl-2-associated X (Bax), cleaved caspase-3 and cleaved caspase-9 protein expression and down-regulation of B-cell lymphoma 2 expression (Yu, Zhang, Ye, et al. 2018).

6.9 Immunomodulatory Peptides

Marine invertebrates are extraordinary resources of immunomodulatory compounds, including peptides, with therapeutic potential to treat human diseases (Natarajan et al. 2016). It is worthy of note that some marine peptides with immunomodulatory properties are multivalent (i.e., with other associated bioactivity such as antioxidant, anticancer and antimicrobial) (Wang et al. 2010; Balseiro et al. 2011; Kang et al. 2019). For instance, from the low molecular weight peptide fraction of Chinese venus *Cyclina sinensis* hydrolysate, the pentadecapeptide RVAPEEHPVEGRYLV was purified, identified and synthesized. This peptide stimulated macrophage phagocytosis by activating the nuclear factor (NF)-κB and the NLRP3 inflammasome signaling pathway, confirming its immunomodulatory effect and utility as an ingredient in functional food or nutraceuticals (Li, Ye, et al. 2019).

6.10 Metal-Chelating/Binding Peptides

A zinc-binding peptide obtained from the Pacific oyster *C. gigas* hydrolysate by plastein reaction (chymotrypsin hydrolysis) has been shown to possess a marked ability to bind zinc ions and significantly enhance zinc bioavailability in the presence of phytic acid. Moreover, the zinc-peptide complex improved the absorption and bioavailability of zinc (Li, Gong, et al. 2019). Similarly, nanocomposite prepared through complexation of *C. rivularis* hydrolysates and zinc improved the zinc bioavailability and solubility in a simulated gastrointestinal digestion (Zhang, Zhou, et al. 2018). Calcium-chelating peptides were obtained from a marine octopus' scrap protein hydrolysate, which promoted calcium uptake in culture cells and was resilient to tannic acid and phytate precipitation and absorptivity inhibition (Wu et al. 2019). These examples illustrate the potential use of metal-chelating/binding peptides from mollusk hydrolysates as supplementary ingredients in functional foods and nutraceuticals to improve the absorptivity and bioavailability of essential metals.

6.11 Osteogenic Peptides and the Nacre Powder

Osteogenic peptides found in and purified from molluscan hydrolysates have significant health benefits to counteract osteoporosis and bone loss, thus, promoting bone formation. For instance, osteogenic bioactivity was observed through the differentiation of mouse mesenchymal stem cells into osteoblasts stimulated by protein hydrolysate of the blue mussel *Mytilus edulis* (Hyung, Ahn, and Je 2018). Then, it was shown that low-molecular weight protein hydrolysate (< 1 kDa) of *M. edulis* upregulated the bone morphogenetic protein-2, followed by activation of osteogenic factors, promoting osteoblast differentiation, which may contribute to bone formation and alleviate osteoporosis symptoms. From protein hydrolysates of the Ark shell *Scapharca subcrenata*, low molecular weight osteogenic peptides were purified that stimulated osteoblast differentiation via mitogen-activated protein kinase (MAPK) and bone morphogenetic protein-2 (BMP-2) pathways. In addition, these Ark shell osteogenic peptides restored femoral bone mineral density and osteoporosis in ovariectomized mice upon daily injection of peptides (Oh, Ahn, Hyung, et al. 2019). Osteogenic bioactivity has also been found in enzymatic hydrolysates of the blue mussel *M. edulis*, as well as in peptides purified from blue mussel and Pacific oyster *C. gigas* protein hydrolysates. Blue mussel protein hydrolysates prepared with proteases (pepsin and trypsin) contained peptides with osteogenic properties that promote osteoblastic growth in culture.

Moreover, this enzymatic hydrolysate proved to have nutritional values (Xu, Zhao, et al. 2019). In addition, a decapeptide purified from the aqueous extract protein of blue mussel, which was hydrolyzed with a neutral protease, stimulated the proliferation of osteoblasts and inhibited the growth of osteoclasts, promoting osteogenesis and preventing osteoporosis, both *in vitro* and *in vivo* models (Xu, Chen, et al. 2019). Similarly, an osteogenic peptide was purified from the *C. gigas* protein digest, which differs from the blue mussel dodecapeptide, that induced pre-osteoblast proliferation in culture through integrin-mediated activation of MAPK pathway (Chen et al. 2019). These examples highlight the particular components, such as peptides, contained in hydrolysate preparations of molluscan tissues that serve as active ingredients in functional foods and nutraceutical formulations.

Another exciting type of molluscan-derived preparation as a source of ingredients for nutraceuticals with osteogenic bioactivity comprises the nacre extracts. Nacre is the lustrous aragonitic inner layer on molluscan shells, known as mother of pearl, of diverse species of mollusks but mainly bivalves. The capacity to stimulate the mineralization of human osteoblasts and bone regeneration has been recognized and reviewed in parallel with other bioactive compounds and extracts from molluscan tissues (Carson and Clarke 2018). Accordingly, the water-soluble extract of nacre has been shown to contain multiple components, like proteins, glycoproteins and polysaccharides, which possess osteogenic properties and play a role in calcium carbonate deposition and mineralization in bone regeneration (Chaturvedi, Singha, and Dey 2013; Brion et al. 2015).

6.12 Peptides Derived from Collagen

Collagen is one of the most abundant proteins and components from the extracellular matrix of animal connective tissues that confer high tensile strength to the structures in which it occurs. Collagen is a generic denomination for triple helix, fibrous proteins that are rich in the uncommon amino acids 4-hydroxiprolyne, and are present in type I collagen and also 5-hydroxilysine. Collagen differs in its amino acid composition and triple-helix structures, as well as plays diverse structural and regulatory roles in the tissues of multicellular organisms. In mollusks, the connective tissues participate in the construction of visceral and skeletal organs, determine the shape of body, protect the animal and bind together muscle tissue that permit locomotion; in cephalopods, cartilaginous tissues constitute the perineural skeleton and scattered segments in the neck, mantle, gills and eye wall; the byssus of bivalves mollusks is a type of extracorporeal "soluble" collagen by which the animals adhere to surfaces (Bairati 1985). Marine collagen and gelatin (denatured collagen), mainly from fish skin, have been obtained for application in tissue engineering, wound healing devices, cosmeceuticals, dietary supplements and nutraceuticals (Salvatore et al. 2020). Owing to the biological diversity and adaptability to aquaculture, marine invertebrates have gained attention and, hence, emerged as a biosafe and bio-sustainable source of collagen and collagen products for biomedical uses (Felician et al. 2018; Rahman 2019). From marine mollusks, collagen has been characterized from different molluscan species and their tissues (Melnick 1958; Bairati 1985), like byssus threads, shells, adductor muscles and mantles of bivalves (Pikkarainen et al. 1968; Mizuta et al. 2004; Tabakaeva, Tabakaev, and Piekoszewski 2018), and cartilage, muscle and skin of cephalopods (Sivakumar, Suguna, and Chandrakasan 2003; Dai et al. 2018; Takema and Kimura 1982; Kimura, Takema, and Kubota 1981).

Peptides derived from collagen and/or gelatin hydrolysates, apart from their nutritional composition, have biological activities and regulatory roles that have been shown to promote health and alleviate disease symptoms caused by chronic disorders. For example, collagen-derived peptides with antioxidant, anti-hypertensive, antimicrobial, anticoagulant, anticancer, anti-glycemic, anti-lipidemic, immunomodulatory, metal-chelating/absorbing and osteogenic activities, in different stages of clinical trials and pharmaceutical development, have been documented (Senevirathne and Kim 2012; Felician et al. 2018; Salvatore et al. 2020). In fact, some of the aforementioned biological activities are contained in peptides derived from molluscan collagen, gelatin and collagen-containing hydrolysates. For instance, short peptides (< 5 kDa) obtained from hydrolyzed collagen of the squid *Dosidicus gigas* by-products possess higher antioxidant and antimutagenic activities (Suárez-Jiménez et al. 2019). From the tunic and skin gelatin hydrolysates of several squid species, antimicrobial and antioxidant peptides (free radical quenching) peptides, in the range of 1–10 kDa, have been characterized (Gómez-Guillén et al. 2011).

Moreover, a non-immunogenic, type II collagen was purified from the cartilage of the Peru Jumbo Flying Squid *Dosidicus gigas* with the capacity of reducing pro-inflammatory mediators and relieving osteoarthritis degenerative symptoms (Dai et al. 2018). Notably, bioactive, food-derived collagen peptides with the properties of, for instance, enhancing the growth of fibroblasts and the synthesis of hyaluronic acid, can be detected in the human blood, at micromolar concentration, after ingestion of collagen hydrolysates, which could explain part of their health-promoting effects (Sato 2017). Thus far, bivalve and cephalopod mollusks figure as an alternative, and sustainable aquaculture source of collagen for the preparation of hydrolysates, bioactive peptides, functional foods and nutraceuticals.

6.13 Cephalopod Ink

Cephalopod mollusks, such as sea hares, cuttlefish, squid and octopus, release ink used as a chemical deterrence and sensory disruption when threatened by predators. Cephalopod ink is a mixture of substances including a black ink containing mainly melanin and a mucous substance which are separately produced in two glands (respectively, the ink sac with its ink gland and the funnel organ with the mucus-producing gland). Eumelanin, a polymer of 5,6-dihydroxyindole (DHI) and 5,6-dihodroxyindole-2-carboxylic (DHICA) acid, which are derived from tyrosine and are in the form of granules, is the primary melanin type found in the ink of cephalopods (Derby 2014). In addition to eumelanin polymers, the cephalopod ink contains diverse melanin-related compounds that include tyrosine, catecholamines (dopamine and DOPA) and enzymes related to melanin biosynthesis, such as tyrosinases (tyrosine hydroxylase or DOPA oxidase), peroxidases and dopachrome-rearranging enzymes. Moreover, in the ink of several species of squid, peptidoglycans that are rich in fucose (6-deoxy-L-galacto-hexopyranose) have been found, as well high levels of metals, like cadmium and copper (Derby 2014). A variable level of proteins also makes up the ink of cephalopods, as well as millimolar concentrations of total free amino acids and ammonium, of which the highest level includes taurine, aspartic acid, glutamic acid, alanine and lysine (Derby et al. 2007). Cephalopod ink has found uses in present Mediterranean cuisine and for millennia in traditional medicine, particularly in China, and, due to its health-beneficial properties, it has attracted attention of researchers and industry that are interested in drug development from natural products and natural food colorants. For instance, the ink extract of the cuttlefish *Sepia officinalis* was effective as an anti-neoplastic substance against Ehrlich ascites carcinoma in Swiss albino mice tumor models (Soliman, Fahmy, and El-Abied 2015). Acid mucopolysaccharides (glycosaminoglycans, see later) contained in sepia ink isolated from squids and cuttlefishes have been chemically and structurally characterized, and their biological activities determined. These bioactivities include chemoprevention, anti-neoplastic, chemo-sensitization, and procoagulant and anticoagulant effects, making sepia ink polysaccharides potentially useful as adjuvant agents of chemotherapy (Li, Luo, and Liu 2018).

Interestingly, a natural ink extract from the Japanese spineless cuttlefish *Sepiella japonica* enriched in eumelanin (up to 94%) displayed antioxidant activity and the capacity to regulate the expression of a number of microRNAs involved in the aging process in mice, resulting in the prolongation of cell cycle and delay in the aging progress.

Moreover, the miRNAs that were regulated by *S. japonica* ink extract were connected to the metabolism of sterols and xenobiotics, the regulation of circadian rhythms and biological processes related to control of oxidative stress and damage (Han et al. 2020). Regarding peptides, an oligopeptide with only three amino acid residues (QPK) from *Sepia* ink and a cyclic-stable analogue induced apoptosis in hormone-independent prostate and lung cancer cells through caspase-3 activation and elevation of pro-apoptotic Bax/Bcl-2 ratio (Huang et al. 2012; Zhang et al. 2017). Likewise, a *Sepia* ink dodecapeptide (LKEENKKKKD) was also active against prostate cancer (PC3) cells, inhibiting cancer cell proliferation in a time- and dose-dependent manner. The anticancer effect was apoptosis-dependent, via activation of tumor antigen p 53 and caspase-3, upregulation of apoptosis regulator Bax and downregulation of Bcl-2 (Huang et al. 2017).

6.14 Lipids

Lipids (or fats) comprise a heterogeneous group of substances having in common the property of insolubility in water, but solubility in nonaqueous solvents. The chemical classes of lipids include natural, glycerol-based fats and oils (triacylglycerols), fatty acid esterified with sugars (glycolipids) or phosphorus (phospholipids), sphingosine-based lipids, as well as isoprene-derived compounds (e.g., isoprenoids and sterols) and lipid-soluble vitamins (vitamins A, E, D and K), among others (Gurr et al. 2016). Lipids are found in all organisms and play essential roles in a wide range of biological processes, from energy storage and composition of cellular membranes to signal transduction and regulation of gene expression. Thus, lipids are essential to keep a well-balanced condition of health.

Polyunsaturated or polyenoic fatty acids (PUFA) are long-chain fatty acids with more than one double bond connecting carbon atoms in the molecule. In animals, long-chain PUFAs, namely, arachidonic (all-*cis*-5,8,11,14–20:4, n-6), eicosapentaenoic (EPA; all-*cis*-5,8,11,14,17–20:5, n-3) and docosahexaenoic (DHA; all-*cis*-4,7,10,13,16,19–22:6, n-3) acids are intermediary precursors of eicosanoids (i.e., a group of signaling molecules, such as cyclic prostaglandins, prostacyclins and thromboxanes, and linear leukotrienes and lipoxins, which are involved in distinct physiological process, such as inflammation, cell adhesion, modulation of platelet activity, and regulation of bronchiolar and vascular tone). Both *n*-3 (ω-3) and *n*-6 (ω-6) series of PUFAs, such as linoleic acid (*cis, cis*-9,12–18:2, *n*-6) and α-linolenic acid (all-*cis*-9,12,15–18:3, *n*-3) are, therefore, essential fatty acids that need to be obtained in the diet to be converted into the biologically active arachidonic acid- and DHA/EPA-derived compounds. However, despite ω-3 and ω-6 PUFAs being essential in the diet, the ratio of the ω-6/ω-3 PUFAs is quite relevant, seeing that the metabolic pathway of arachidonic acid (ω-6 PUFA) generates majorly pro-inflammatory compounds, while the metabolic route of DHA and EPA (ω-3 PUFAs) results mainly in anti-inflammatory modulators (Gurr et al. 2016; Tan et al. 2020). In fact, the health-beneficial properties of the ω-3 PUFA DHA and EPA in the diet and dietary supplement have been recognized by the medical community to prevent cardiovascular diseases and inflammatory processes in adults, and promote the development of the brain of the fetus during pregnancy (Harris 2004; AbuMweis et al. 2018). Fish oils are a rich source of PUFAs, notably, with high DHA and EPA content, despite the substantial quantity of monounsaturated fatty acids (MUFAs) (Sahena et al. 2009). From cultivable and invasive Pacific oyster *C. gigas*, several classes of health-promoting lipids were identified and quantified such as high-quality phospholipids (phosphatidylcholine), PUFAs (DHA and EPA), as well as plasmalogens and phytosterols (Dagorn et al. 2016). These molluscan resources constitute an economic lipid supply of marine origin that found applications as food supplements and nutraceuticals. An oily lipid extract from the hard-shelled mussel *Mytilus coruscus*, with a relative abundance of MUFAs and ω-3 and ω-6 PUFAs, prepared through CO_2 supercritical procedure, were shown to improve the clinical conditions of patients with rheumatoid arthritis by enhancing the balance between pro- and anti-inflammatory mediators, indicating its applicability as a nutraceutical and an adjuvant supplement to treat arthritis (Fu et al. 2015). Importantly, due to high market demand for PUFAs, bivalves have emerged as a sustainable source of high quality and beneficial polyenoic ω-3 lipids, mainly because certain mollusk species of commercial interest are amenable to aquaculture (Tan et al. 2020). In fact, green shell mussels (GSM—*Perna canaliculus*) is one of these cases, from which its oil is considered a high-value product that contains lipid components that are not present in available products prepared with regular fish oil (Miller, Pearce, and Bettjeman 2014).

Another interesting class of bioactive lipids found in mollusks comprises the particular class of furan-fatty acids (F-fatty acids). These particular lipids contain an alkyl unit linked to the C-2 of furan ring and a carboxyalkyl chain linked at C-5. F-fatty acids are potent free radical scavengers that protect PUFAs from lipid peroxidation. F-fatty acids were purified from the New Zealand green-lipped mussels *Perna canaliculus* and are one of the active ingredients of the lipid-rich extract Lyprinol used to treat chronic inflammatory diseases like osteoarthritis (Wakimoto et al. 2011; Halpern 2000).

6.15 Sterols and Their Precursor Squalene

Steroids comprise a group of lipids that have a ring system of four fused rings (three with six carbon atoms and one with five), known as 1,2-cyclopentanoperhydrophenanthrene. Sterols are steroids which have a variable number of hydroxyls (polar head groups) and an alkyl side-chain. In eukaryotic cells of animals, cholesterol (cholesteryl alcohol)—a structural lipid—is the predominant sterol which composes and confers rigidity to plasma membrane that is essential for membrane stability. Moreover, cholesterol is the precursor for steroid hormones, such as the corticoids (mineralocorticoids and glucocorticoids) and the sex hormones (progesterone, androgens and estrogens), as well as the bile salts (polar derivatives of cholesterol) and vitamin D hormone.

Fat-soluble vitamins (i.e., vitamins A, D, E and K) are daily required in the diet in micrograms (vitamin A, D and E) and even milligram quantities (vitamin E) to maintain health. Vitamin A, of which *all*-trans-retinol is the parent molecule, is found in fish oil and other animal sources and is essential to avoid preventable blindness. Vitamin D is essential in the diet to sustain bone growth and development, and prevent bone disease rickets in children and osteomalacia in adults. Intake in the diet is required when sunlight exposure is inadequate, because its synthesis that occurs in the skin in response to sunlight is sufficient to attend the body demand. Its active form (hormone) is 1,25-dihydroxycholecalciferol (calcitriol) which increases the efficiency of absorption of dietary calcium and phosphate through enterocytes, as well as promotes osteogenesis and cartilage formation. In view of the fact that these vitamins are lipid-soluble and tend to accumulate in tissues, a well-balanced intake is recommendable to maintain health, avoiding deficiency that impairs diseases and excess that is harmful and detrimental to the liver. Vitamin E (tocopherol) is a lipid-soluble antioxidant that prevents peroxidation of long-chain PUFAs, which otherwise destroy the membrane of cells if not counteracted. An anti-inflammatory effect is also attributed to vitamin E because of its inference with the synthesis of pro-inflammatory eicosanoids (prostaglandins). Vitamin K comprises a group of naphthoquinone compounds with different side-chain substituents (e.g., phylloquinones, vitamin K1 and menaquinones, vitamin K2). Vitamin K is essential for adequate blood coagulation, but vitamin K deficiency is rare (Gurr et al. 2016). Mollusks are excellent sources of lipid-soluble vitamins and PUFAs, as observed in native and farmed mussels (Merdzhanova et al. 2014; Chakraborty et al. 2016; Stancheva, Merdzhanova, and Dobreva 2017).

The presence of variable levels of a variety of non-cholesterol sterols, which include the anti-hypercholesterolemic phytosterols (i.e., sitosterol, campesterol, stigmasterol, brassicasterol, Δ5-avenasterol, sitostanol, campestanol), makes marine mollusks an important source of lipids with health-promoting effects (Phillips et al. 2012). Aqueous extract from the edible part of the freshwater clam *Corbicula fluminea* contains, among other substances, the phytosterols campesterol, brassicasterol, stigmasterol and β-sitosterol, which were shown to contribute to the decrease of cholesterol, blood levels, reduction of accumulation of hepatic lipids, and the increase of cholesterol catabolism to bile salts, in an experimental mouse model of hypercholesterolemia (Chijimatsu et al. 2009). Moreover, the organic extract of this freshwater clam, which contains phytosterols, similarly reduced the hepatic cholesterol levels and enhanced the cholesterol catabolism in the mouse model of nonalcoholic steatohepatitis. However, the improved hepatoprotection and steatohepatitis resolution effects were obtained with a total aqueous extract that usually contains a high proportion of glycine (and proteins) (Yao et al. 2018).

In the biosynthesis of steroids, the successive enzymatic condensations of isoprene units lead to the formation of squalene, a 30-carbon (C-30) intermediary compound (presterol isoprenoid) that is converted in 2,3-epoxysqualene and, subsequently, cyclized to animal and plant sterols. Owing to its structure, a lipophilic triterpene molecule with six double bonds, squalene is an efficient antioxidant that acts as scavenger and quencher of some types of free radicals, detaining lipid peroxidation and contributing to protection of the skin and cardiovascular system. In addition, squalene has beneficial effects on dyslipidemia (Micera et al. 2020). Moreover, the dietary intake of squalene has been implicated in decreasing the risk for several types of cancer, due to the chemoprotective properties of squalene that appears to control carcinogenesis by several molecular mechanisms (Smith 2000). In fact, the selective chemoprotective activity of squalene have been demonstrated *in vitro*, by comparing the effects on healthy bone marrow and neuroblastoma cells (Das et al. 2003), and *in vivo* in drug-induced genotoxicity in mice (Narayan et al. 2010) and doxorubicin-treated allograft mice (Narayan Bhilwade et al. 2019). These studies indicated that squalene is a potential adjuvant in chemotherapy that acts by inhibiting inflammation and protecting healthy cells in detriment to tumor cells. Squalene was first isolated from the liver oil of sharks (*Squalus milsukurii* and other species) more than one hundred years ago, and further from plant extracts (Lozano et al. 2018). Squalene was also found in mollusk tissues, such as in the edible abalone *Haliotis gurneri* and the mussel *Mytilus edulis* (Teshima and Kanazawa 1974). Despite squalene being found in abalone and mussel species, continuous studies are required to confirm the distribution and the capacity of squalene biosynthesis in mollusks. Regardless, bivalves contain specific mixtures of C26 to C29 sterols that include 24-methyl and 24-ethyl sterols, and unusual *cis, cis*-5,7-sterols and 4a-methyl sterols, as particular characteristics (Urich 1994). Additionally, the biosynthesis of steroids from mevalonate and acetate was observed in the cephalopods *Sepia officinalis* and *Octopus vulgaris* (Voogt 1973). In Table 6.3, different classes of lipids with beneficial effects to human health found in mollusks are presented. Based on these examples, lipids from mollusks are prominent active ingredients for functional foods and nutraceuticals that can be obtained on a commercial scale and at sustainable levels for health benefits.

Table 6.3 Some Lipids of Importance for Health from Mollusks

Lipid	Function	Health Benefits	Species	Reference
PUFAs linoleic acid (18:2, ω-6) α-linolenic acid (18:3, ω-3) eicosapentaenoic acid, EPA (20:5, ω-3) docosahexaenoic acid, DHA (22:6, ω-3)	Precursors of signaling molecules involved in pain, inflammation and chemoattraction; inhibition of both the 5-lipoxygenase and cyclooxygenase; arachidonate oxygenation pathways	Prevent cardiovascular diseases and inflammatory processes; contribute with neural structure and function; anti-inflammatory	Diverse species of edible mollusks	(Tan et al. 2020) (Dagorn et al. 2016) (Murphy, Mann, and Sinclair 2003; Murphy et al. 2002)

Table 6.3 Some Lipids of Importance for Health from Mollusks. Continued

Lipid	Function	Health Benefits	Species	Reference
Phytosterols sitosterol, campesterol, stigmasterol, brassicasterol, Δ5-avenasterol, sitostanol, campestanol	Lipid-lowering effect; inhibition of absorption of both endogenous and exogenous cholesterol; reduction of serum total cholesterol levels	Anticholesterolemic; reduction of accumulation of hepatic lipids; hepatoprotection	Diverse species of edible mollusks	(Phillips et al. 2012) (Chijimatsu et al. 2009; Chijimatsu et al. 2015) (Yao et al. 2018)
Squalene	A 30-carbon intermediary compound (presterol isoprenoid) precursor of sterols; free radical scavenger	Decreasing the risk for several types of cancer; chemoprotective; antidislipidemic	Abalone *Haliotis gurneri* Blue mussel *Mytilus edulis*	(Teshima and Kanazawa 1974)
Furan-fatty acids	Protection of PUFAs from lipid peroxidation; free radical scavenger	Anti-inflammatory antioxidant; anti-arthritis	NZ green-lipped mussel, *Perna canaliculus*	(Wakimoto et al. 2011) (Halpern 2000)
Fat-soluble vitamins A, E, D and K	Precursors of visual pigments and hormones; antioxidant; prevents lipid peroxidation of PUFAS; sustains bone growth and development; adequate blood coagulation	Avoid preventable blindness; bone growth and development; antioxidant; blood coagulation	Diverse species of edible mollusks	(Stancheva, Merdzhanova, and Dobreva 2017)

6.16 Polysaccharides and Complex Glycans

Carbohydrates (polyhydroxy aldehydes or ketones) are the most abundant biomolecules in nature and are classified in mono-, oligo- or polysaccharides. Examples of mono- and oligosaccharides are glucose and sucrose (a disaccharide), respectively. Polysaccharides are linear or branched polymers composed by hundreds to thousands of sugar residues (monomers), with medium to high molecular weight. Polysaccharides (glycans) are divided into homopolysaccharides and heteropolysaccharides, if they contain only a single type of covalently linked monomeric entity, or if more than one kind of monosaccharide is joined together in the polymeric chains, respectively. Some homopolysaccharides play a role in the storage of energy (e.g., glycogen and starch), while others serve as structural material (e.g., cellulose, β1→4, glucose [Glc] and chitin, β1→4, N-acetylglucosamine [GlcNAc]). Heteropolysaccharides provide extracellular structural support, as in the cell envelope of bacteria (in peptidoglycan) and in the cell wall of algae (agar), as well as in the extracellular matrix (glycosaminoglycans) of animal tissues that holds individual cells together and provides a gel-like substrate for diffusion of substances and cells.

Glycosaminoglycans (GAGs), also known as mucopolysaccharides, comprise a family of complex, linear, negatively charged polymers composed of repeating disaccharide units in which one of the two monosaccharides is invariably either N-acetylglucosamine (GlcNAc) or N-acetylgalactosamine (GalNAc); the other unit is a uronic acid derivative (Nelson and Cox 2012). GAGs, such as heparin/heparan sulfate glycosaminoglycan and chondroitin/dermatan sulfate glycosaminoglycan, interact with numerous proteins in the extracellular matrix and modulate their activity, thus playing essential roles in critical biological processes such as signaling and regulation of cell development, angiogenesis, axonal growth, cancer progression, microbial pathogenesis and anticoagulation (Sasisekharan, Raman, and Prabhakar 2006). Chondroitin sulfate, a glycosaminoglycan found in the cartilage and extracellular matrix, takes part in the regulation of cell development, cell adhesion, proliferation and differentiation, which promotes and accelerates the regeneration of damaged structures in osteoarthritis (Bishnoi et al. 2016). Gulati and Poluri (Gulati and Poluri 2016) summarize some of the physiological functions and pathophysiological effects in which GAGs are involved. For example, the physiological functions of heparin and heparan sulfate include their participation in cellular growth, cell-cell and cell-matrix adhesion, inflammation, angiogenesis, localization of cytokines and chemokines. Dermatan sulfate confers tensile strength to the skin and plays an essential role in the stability of extracellular matrix, blood coagulation, cellular proliferation, organogenesis and wound healing. In turn, chondroitin and chondroitin sulfate are biomolecules that control cellular processes, which include the inhibition of axonal growth, cell migration and plasticity, modulation of neural stem cells and cartilage repair. Keratan sulfate is responsible for maintaining the hydration level and transparency in the cornea, controlling inflammation, neural regeneration and plasticity, as well as decelerating cartilage damage. Hyaluronic acid, physiologically, controls water homeostasis, lubrication, structural integrity, sequestration of free radicals and plasma protein distribution, regulates inflammation and cell signaling events (proliferation, migration and differentiation). Based on these physiological functions a number of GAGs-based drugs and analogues are in the market to treat several diseases related to GAGs disfunctions, like from thromboembolism and viral hepatitis to osteoarthritis and venous and arterial leg diseases, among others (Gulati and Poluri 2016). Hence, complex glycans have multiple benefits to human health to circumvent glycan-related diseases, and GAGs-containing supplements and nutraceuticals arise as valuable products to maintain health and alleviate disease symptoms.

Indeed, a large diversity of GAG structures with high chemical heterogeneity is found in nature, owing to the complexity and template-independence of GAG biosynthesis. In this respect, marine organisms have been recognized as rich sources of biologically active GAGs that are useful for nutraceutical, cosmeceutical and pharmacological applications (Ruocco et al. 2016; Pomin 2014). Glycan-based nutraceuticals from marine organisms, like their terrestrial counterparts, are majorly sulfated and the level of sulfation is critical for the biological activity of GAGs and the maintenance of health (Tang et al. 2019; Sasisekharan, Raman, and Prabhakar 2006). Although most commercial polysaccharides of marine origin are from algae (e.g., alginate, carrageenan and fucoidan), a range of structural and physicochemical diverse polysaccharides can be extracted and purified from mollusks, particularly from species of bivalves, gastropods and cephalopods. These polysaccharides include neutral and sulfated polysaccharides, GAGs (e.g., heparin/heparan sulfate, chondroitin/chondroitin sulfate, dermatan sulfate, keratan sulfate, hyaluronan, etc.) and heterogeneous glycans with undefined pattern of repeating units. Extracts and (medicinal) preparations from mollusks that contain bioactive polysaccharides have been reported. These molluscan-derived preparations have many health-promoting and disease-preventing properties, like antiatherogenic, anticoagulant and antithrombotic, antioxidant, antitumor, antiviral, and immunomodulatory (Wang et al. 2019). For instance, from the tissues of *S. officinalis*, like cartilage, cornea, integument, ink and ink sac, a complex spectrum of chondroitin and chondroitin sulfate compounds have been characterized. From bivalves, heparin and heparan sulfates are often found; despite these, GAGs are not restricted to this class of mollusks. Polysaccharides and complex glycans in mollusks differ in content, chemical structure and composition, as well as vary with species, tissue, developmental stages and habitat, among other variables (Wang et al. 2019). The heterogeneity of polysaccharide found in mollusks reinforces the quality of these organisms as a rich source of bioactive glycans. For instance, total GAGs isolated from the entire body tissue of the freshwater mollusk bivalve *Anodonta* are composed of chondroitin sulfate, nonsulfated chondroitin and heparin in different proportions (Volpi and Maccari 2005).

Furthermore, GAGs from marine organisms differ from their terrestrial counterparts regarding the chemical composition, chain length, molecular weight and, importantly, the grade, position and patterns of sulfation, which influence their physiological effects and therapeutic applicability (Valcarcel et al. 2017). These molluscan sulfated glycosaminoglycans such as chondroitin sulfate possess antiviral and antimetastatic activities and improve cartilage constitution, whereas, dermatan sulfate and keratan sulfates modulate neurite outgrowth, serving for nerve regeneration (Higashi et al. 2016; Valcarcel et al. 2017). Some examples of biologically and pharmacologically active polysaccharides from mollusks that useful as nutraceuticals are listed in Table 6.4. Accordingly, GAGs that have been purified represent only a fraction of the structural and functional diversity of polysaccharides found in mollusks.

Table 6.4 Examples of Biological and Pharmacological Polysaccharides from Mollusks

Biological Activity and Polysaccharide Family	Species and Common Name	Reference
Anticoagulant		
Heparin-like	Baby clam, *Katelysia opima*	(Vijayabaskar, Balasubramanian, and Somasundaram 2008)
Heparin-like	Marine clam, *Meretrix meretrix*	(Saravanan, Vairamani, and Shanmugam 2010)
Heparin	Backwater clam, *Donax cuneatus*	(Vijayabaskar and Somasundaram 2012)
Antithrombotic		
Dermatan sulfate	Marine clam, *Scapharca inaequivalvis*	(Volpi and Maccari 2008)
heparan sulfate-like	Lion's paw scallop, *Nodipecten nodosus*	(Gomes et al. 2010)
Antimicrobial, Antiviral and Anti-Parasite		
Heparin and heparin-like	Bobtail squid, *Euprymna berryi*	(Shanmugam et al. 2008)
Sulfated chitosan	Cuttlefish, *Sepia pharaonis*	(Karthik et al. 2016)
Heparan sulfate-like	Lion's paw scallop, *Nodipecten nodosus*	(Bastos et al. 2019)
Antioxidant and Immunomodulatory		
Complex glycans	Marine clam, *Meretrix meretrix*	(Wang et al. 2018; Li et al. 2015)
Xylated glycosaminoglycan	Marine gastropod, *Babylonia spirata*	(Chakraborty and Salas 2020)
Antiproliferative (Antitumoral)		
Heparin	Giant clam, *Tridacna maxima,* green mussel, *Perna viridis*	(Muthuvel et al. 2009)
Chondroitin sulfate	Cuttlefish, *Sepia pharaonis*	(Seedevi et al. 2017)
Heparan sulphate/heparin-like	Cockle, *Cerastoderma edule*	(Aldairi, Ogundipe, and Pye 2018)
Fibrinolytic		
Heparin-like	Clam, *Coelomactra antiquata*	(Du et al. 2019)
Metal-Binding and Transport		
Heparan sulphate/heparin-like	Freshwater mussel, *Anodonta californiensis*	(Hovingh and Linker 1993)
Osteogenic, Proliferative and Cell Modulator		
Keratan sulfate	Clam, *Mactra chinensis*	(Higashi et al. 2016)

Note: Chondroitin sulfates consist of repeating units of GalNAc-GlcA disaccharide joined by β1,4 and β1,3 linkages. Dermatan sulfates derive from Chondroitin sulfates by enzymatic 5-epimerization of some GlcA residues to IdoA. Keratan sulfates are composed of repeating disaccharide units of Gal and GlcNAc joined by β1,4 and β1,3 linkages. Heparins and heparan sulfates are constituted of alternating α1,4 GlcNAc and β1,4 GlcA units. Chitin is a linear homopolysaccharide composed of GalNAc residues in (β1,4) linkage. Chitosan is the deacetylated derivative of chitin. GalNAc: N-acetylgalactosamine; GlcA: glucuronic acid; IdoA: L-iduronic acid; Gal: galactose.

Table 6.5 Examples of Nutraceuticals Derived from Mollusks in Experimental and Commercial Stages

Common Name	Scientific Name	Active Ingredient	Form	Bioactivity	Reference
Experimental					
Pacific oyster	*Crassostrea gigas*	Fermented extract	Lyophilized powder	Bone loss suppression, anti-osteoclastogenic; bone mineralization	(Ihn et al. 2019; Molagoda et al. 2019)
Pacific oyster	*Crassostrea gigas*	Pentapeptide	Protein hydrolysate	Suppresses wrinkle formation (anti-photoaging)	(Bang, Jin, and Choung 2020)
Pacific oyster	*Crassostrea gigas*	LMW peptides	Lyophilized powder	Immunostimulant effects; antitumor	(Wang et al. 2010)
Clam	*Mactra veneriformis*	Polysaccharides	Water-soluble extract	Hepatoprotection, antioxidant	(Wang et al. 2020)
Green lip abalone	*Haliotis laevigata*	LMW compounds mix	Non-lipophilic extract	Antiviral	(Dang, Benkendorff, and Speck 2011)
Mollusk	*Arca subcrenata*	Polypeptides	Protein fraction	Tumor growth inhibition	(Hu et al. 2012)
Common octopus	*Octopus vulgaris*	Peptides	Protein hydrolysate	Antioxidant	(Ben Slama-Ben Salem et al. 2017)
Golden cuttlefish	*Sepia esculenta*	Tripeptide	Ink extract/synthetic peptide	Antitumoral; induction of apoptosis	(Zhang et al. 2017; Huang et al. 2012)
King scallop	*Pecten maximus*	Cytokine-like; growth factors	Water soluble matrix	Chondrogenic effects; redifferentiation of chondrocytes	(Bouyoucef et al. 2018)
Commercial					
Green-lipped mussel	*Perna canaliculus*	Lipid-rich extract	Powder (Lyprinol)	Anti-inflammatory (anti-osteoarthritis)	(Halpern 2000)
Asian green mussel	*Perna viridis*	Lyophilized extract	Cadalmin (GMe)	Reliefs of chronic arthritis pain; resolution of inflammatory disorders	(Chakraborty 2012)
Green-lipped mussel	*Perna canaliculus*	Lipid-rich extract (Lyprinol)	Seatone	Regeneration of arthritic and injured joints; relieves arthritis symptoms	(Cobb and Ernst 2006)
Med. mussel	*M. galloprovincialis*	Lipoprotein extract	E-MHK-0103 (Mineraxin™)	Alleviates menopausal symptoms	(Corzo et al. 2017)

6.17 Concluding Remarks

As one can conclude from these examples presented in this chapter, bioactive substances derived from mollusks with health-promoting and disease-alleviating properties are diverse among species and (bio-)chemical families. These biomolecules can be utilized either as a blend of compounds of distinct chemical classes, like in hydrolysates, preparations enriched in one class of compounds (e.g., bioactive lipids, complex carbohydrates, protein fragments), or wholly purified entities (e.g., peptides) (Table 6.5). Purified molluscan substances, extracts and hydrolysates all provide invaluable biologically and pharmacologically active ingredients to prepare a range of formulations and functional food and nutraceutical products. The global market for natural products, functional foods, cosmeceuticals and nutraceuticals, in particular, is expanding worldwide because of the societal search for a healthier and nature-integrated lifestyle. The advanced knowledge gathered with the advent of novel analytical technologies confirm, in general, the efficacy of active ingredients from traditional and folk medicine. These cutting-edge technologies have revealed, at the molecular level and at a chemical basis, the beneficial health effects of the active ingredients in extracts and preparations derived from mollusks. Moreover, the uncountable number of molluscan species existing in nature and the potential adaptability of several species to aquaculture make these organisms biosustainable resources of commodities for the functional food, nutraceutical and health industries.

References

AbuMweis, S., S. Jew, R. Tayyem, and L. Agraib. 2018. "Eicosapentaenoic acid and docosahexaenoic acid containing supplements modulate risk factors for cardiovascular disease: a meta-analysis of randomised placebo-control human clinical trials." *J Hum Nutr Diet* 31 (1):67–84. doi: 10.1111/jhn.12493.

Ahmad, Tarek B., Lei Liu, Michael Kotiw, and Kirsten Benkendorff. 2018. "Review of anti-inflammatory, immune-modulatory and wound healing properties of molluscs." *Journal of Ethnopharmacology* 210:156–178. doi: https://doi.org/10.1016/j.jep.2017.08.008.

Aldairi, Abdullah Faisal, Olanrewaju Dorcas Ogundipe, and David Alexander Pye. 2018. "Antiproliferative activity of glycosaminoglycan-like polysaccharides derived from marine molluscs." *Mar Drugs* 16 (2):63. doi: 10.3390/md16020063.

Bairati, Aurelio. 1985. "The collagens of the mollusca." In *Biology of Invertebrate and Lower Vertebrate Collagens*, edited by A. Bairati and R. Garrone, 277–297. Boston, MA: Springer US.

Balseiro, Pablo, Alberto Falcó, Alejandro Romero, Sonia Dios, Alicia Martínez-López, Antonio Figueras, Amparo Estepa, and Beatriz Novoa. 2011. "Mytilus galloprovincialis myticin C: A chemotactic molecule with antiviral activity and immunoregulatory properties." PLOS ONE 6 (8):e23140. doi: 10.1371/journal.pone.0023140.

Bang, J. S., Y. J. Jin, and S. Y. Choung. 2020. "Low molecular polypeptide from oyster hydrolysate recovers photoaging in SKH-1 hairless mice." *Toxicol Appl Pharmacol* 386:114844. doi: 10.1016/j.taap.2019.114844.

Bastos, M. F., L. Albrecht, A. M. Gomes, S. C. Lopes, C. P. Vicente, R. P. de Almeida, G. C. Cassiano, R. J. Fonseca, C. C. Werneck, M. S. Pavão, and F. T. Costa. 2019. "A new heparan sulfate from the mollusk Nodipecten nodosus inhibits merozoite invasion and disrupts rosetting and cytoadherence of Plasmodium falciparum." *Mem Inst Oswaldo Cruz* 114:e190088. doi: 10.1590/0074-02760190088.

Ben Slama-Ben Salem, R., I. Bkhairia, O. Abdelhedi, and M. Nasri. 2017. "Octopus vulgaris protein hydrolysates: characterization, antioxidant and functional properties." *J Food Sci Technol* 54 (6):1442–1454. doi: 10.1007/s13197-017-2567-y.

Ben Slama-Ben Salem, R., N. Ktari, I. Bkhairia, R. Nasri, L. Mora, R. Kallel, S. Hamdi, K. Jamoussi, T. Boudaouara, A. El-Feki, F. Toldrá, and M. Nasri. 2018. "In vitro and in vivo anti-diabetic and anti-hyperlipidemic effects of protein hydrolysates from Octopus vulgaris in alloxanic rats." *Food Res Int* 106:952–963. doi: 10.1016/j.foodres.2018.01.068.

Benoist, Louis, Baptiste Houyvet, Joël Henry, Erwan Corre, Bruno Zanuttini, and Céline Zatylny-Gaudin. 2020. "In-depth in silico search for cuttlefish (sepia officinalis) antimicrobial peptides following bacterial challenge of haemocytes." *Marine Drugs* 18 (9):439.

Bishnoi, M., A. Jain, P. Hurkat, and S. K. Jain. 2016. "Chondroitin sulphate: a focus on osteoarthritis." *Glycoconj J* 33 (5):693–705. doi: 10.1007/s10719-016-9665-3.

Bouyoucef, M., R. Rakic, T. Gómez-Leduc, T. Latire, F. Marin, S. Leclercq, F. Carreiras, A. Serpentini, J. M. Lebel, P. Galéra, and F. Legendre. 2018. "Regulation of extracellular matrix synthesis by shell extracts from the marine bivalve pecten maximus in human articular chondrocytes-application for cartilage engineering." *Mar Biotechnol (NY)* 20 (4):436–450. doi: 10.1007/s10126-018-9807-7.

Brion, A., G. Zhang, M. Dossot, V. Moby, D. Dumas, S. Hupont, M. H. Piet, A. Bianchi, D. Mainard, L. Galois, P. Gillet, and M. Rousseau. 2015. "Nacre extract restores the mineralization capacity of subchondral osteoarthritis osteoblasts." *J Struct Biol* 192 (3):500–509. doi: 10.1016/j.jsb.2015.10.012.

Carson, M. A., and S. A. Clarke. 2018. "Bioactive compounds from marine organisms: Potential for bone growth and healing." *Mar Drugs* 16 (9). doi: 10.3390/md16090340.

Chai, Tsun-Thai, Yew-Chye Law, Fai-Chu Wong, and Se-Kwon Kim. 2017. "Enzyme-assisted discovery of antioxidant peptides from edible marine invertebrates: a review." *Mar Drugs* 15 (2):42.

Chakraborty, K. 2012. "Green Mussel extract (GMe) goes commercial: First nutraceutical produced by an ICAR Institute." *CMFRI Newsletter* 135 (135):5–6.

Chakraborty, Kajal, Selsa J. Chakkalakal, Deepu Joseph, P. K. Asokan, and K. K. Vijayan. 2016. "Nutritional and antioxidative attributes of green mussel (Perna viridis L.) from the Southwestern Coast of India." *Journal of Aquatic Food Product Technology* 25 (7):968–985. doi: 10.1080/10498850.2015.1004498.

Chakraborty, Kajal, and Soumya Salas. 2020. "First report of a glycosaminoglycan-xylopyranan from the buccinid gastropod mollusk Babylonia spirata attenuating proinflammatory 5-lipoxygenase." *J Food Biochemist* 44 (1):e13082. doi: https://doi.org/10.1111/jfbc.13082.

Chalamaiah, M., W. Yu, and J. Wu. 2018. "Immunomodulatory and anticancer protein hydrolysates (peptides) from food proteins: A review." *Food Chem* 245:205–222. doi: 10.1016/j.foodchem.2017.10.087.

Charlet, M., S. Chernysh, H. Philippe, C. Hetru, J. A. Hoffmann, and P. Bulet. 1996. "Innate immunity. Isolation of several cysteine-rich antimicrobial peptides from the blood of a mollusc, Mytilus edulis." *J Biol Chem* 271 (36):21808–13. doi: 10.1074/jbc.271.36.21808.

Chatterjee, R., T. K. Dey, A. Roychoudhury, D. Paul, and P. Dhar. 2020. "Enzymatically excised oligopeptides from Bellamya bengalensis shows potent antioxidative and anti-hypertensive activity." *J Food Sci Technol* 57 (7):2586–2601. doi: 10.1007/s13197-020-04295-8.

Chaturvedi, R., P. K. Singha, and S. Dey. 2013. "Water soluble bioactives of nacre mediate antioxidant activity and osteoblast differentiation." *PLoS One* 8 (12):e84584. doi: 10.1371/journal.pone.0084584.

Chen, Da, Zunying Liu, Wenqian Huang, Yuanhui Zhao, Shiyuan Dong, and Mingyong Zeng. 2013. "Purification and characterisation of a zinc-binding peptide from oyster protein hydrolysate." *J Funct Foods* 5 (2):689–697. doi: https://doi.org/10.1016/j.jff.2013.01.012.

Chen, H., Z. Xu, F. Fan, P. Shi, M. Tu, Z. Wang, and M. Du. 2019. "Identification and mechanism evaluation of a novel osteogenesis promoting peptide from Tubulin Alpha-1C chain in Crassostrea gigas." *Food Chem* 272:751–757. doi: 10.1016/j.foodchem.2018.07.063.

Cheng, S., M. Tu, H. Chen, Z. Xu, Z. Wang, H. Liu, G. Zhao, B. Zhu, and M. Du. 2018. "Identification and inhibitory activity against α-thrombin of a novel anticoagulant peptide derived from oyster (Crassostrea gigas) protein." *Food Funct* 9 (12):6391–6400. doi: 10.1039/c8fo01635f.

Cheng, S., M. Tu, H. Liu, Y. An, M. Du, and B. Zhu. 2021. "A novel heptapeptide derived from Crassostrea gigas shows anticoagulant activity by targeting for thrombin active domain." *Food Chem* 334:127507. doi: 10.1016/j.foodchem.2020.127507.

Cheong, S. H., E. K. Kim, J. W. Hwang, Y. S. Kim, J. S. Lee, S. H. Moon, B. T. Jeon, and P. J. Park. 2013. "Purification of a novel peptide derived from a shellfish, Crassostrea gigas, and evaluation of its anticancer property." *J Agric Food Chem* 61 (47):11442–6. doi: 10.1021/jf4032553.

Chijimatsu, T., I. Tatsuguchi, H. Oda, and S. Mochizuki. 2009. "A Freshwater clam (Corbicula fluminea) extract reduces cholesterol level and hepatic lipids in normal rats and xenobiotics-induced hypercholesterolemic rats." *J Agric Food Chem* 57 (8):3108–12. doi: 10.1021/jf803308h.

Chijimatsu, T., M. Umeki, S. Kobayashi, Y. Kataoka, K. Yamada, H. Oda, and S. Mochizuki. 2015. "Dietary freshwater clam (Corbicula fluminea) extract suppresses accumulation of hepatic lipids and increases in serum cholesterol and aminotransferase activities induced by dietary chloretone in rats." *Biosci Biotechnol Biochem* 79 (7):1155–63. doi: 10.1080/09168451.2015.1012147.

Cobb, Christopher S., and Edzard Ernst. 2006. "Systematic review of a marine nutriceutical supplement in clinical trials for arthritis: the effectiveness of the New Zealand green-lipped mussel Perna canaliculus." *Clinical Rheumatology* 25 (3):275–284. doi: 10.1007/s10067-005-0001-8.

Correia-da-Silva, M., E. Sousa, M. M. M. Pinto, and A. Kijjoa. 2017. "Anticancer and cancer preventive compounds from edible marine organisms." *Semin Cancer Biol* 46:55–64. doi: 10.1016/j.semcancer.2017.03.011.

Corzo, L., S. Rodriguez, R. Alejo, L. Fernandez-Novoa, G. Aliev, and R. Cacabelos. 2017. "E-MHK-0103 (Mineraxin™): A novel nutraceutical with biological properties in menopausal conditions." *Curr Drug Metab* 18 (1):39–49. doi: 10.2174/1389200217666161014151341.

Dagorn, F., A. Couzinet-Mossion, M. Kendel, P. G. Beninger, V. Rabesaotra, G. Barnathan, and G. Wielgosz-Collin. 2016. "Exploitable lipids and fatty acids in the invasive oyster crassostrea gigas on the French Atlantic coast." *Mar Drugs* 14 (6). doi: 10.3390/md14060104.

Dai, M., X. Liu, N. Wang, and J. Sun. 2018. "Squid type II collagen as a novel biomaterial: Isolation, characterization, immunogenicity and relieving effect on degenerative osteoarthritis via inhibiting STAT1 signaling in pro-inflammatory macrophages." *Mater Sci Eng C Mater Biol Appl* 89:283–294. doi: 10.1016/j.msec.2018.04.021.

Dang, V. T., K. Benkendorff, and P. Speck. 2011. "In vitro antiviral activity against herpes simplex virus in the abalone Haliotis laevigata." *J Gen Virol* 92 (Pt 3):627–37. doi: 10.1099/vir.0.025247-0.

Das, B., H. Yeger, H. Baruchel, M. H. Freedman, G. Koren, and S. Baruchel. 2003. "In vitro cytoprotective activity of squalene on a bone marrow versus neuroblastoma model of cisplatin-induced toxicity: implications in cancer chemotherapy." *Euro J Canc* 39 (17):2556–2565. doi: 10.1016/j.ejca.2003.07.002.

De Zoysa, M. 2012. "Medicinal benefits of marine invertebrates: sources for discovering natural drug candidates." *Adv Food Nutr Res* 65:153–69. doi: 10.1016/b978-0-12-416003-3.00009-3.

De Zoysa, M., C. Nikapitiya, I. Whang, J. S. Lee, and J. Lee. 2009. "Abhisin: a potential antimicrobial peptide derived from histone H2A of disk abalone (Haliotis discus discus)." *Fish Shellfish Immunol* 27 (5):639–46. doi: 10.1016/j.fsi.2009.08.007.

Derby, C. D. 2014. "Cephalopod ink: production, chemistry, functions and applications." *Mar Drugs* 12 (5):2700–30. doi: 10.3390/md12052700.

Derby, C. D., C. E. Kicklighter, P. M. Johnson, and X. Zhang. 2007. "Chemical composition of inks of diverse marine molluscs suggests convergent chemical defenses." *J Chem Ecol* 33 (5):1105–13. doi: 10.1007/s10886-007-9279-0.

Du, Z., X. Jia, J. Chen, S. Zhou, J. Chen, X. Liu, X. Cao, S. Zhong, and P. Hong. 2019. "Isolation and characterization of a heparin-like compound with potent anticoagulant and fibrinolytic activity from the clam coelomactra antiquata." *Mar Drugs* 18 (1). doi: 10.3390/md18010006.

FAO/INFOODS. 2016. "FAO/INFOODS global food composition database for fish and shellfish version 1.0-uFiSh1.0."

Felician, F. F., C. Xia, W. Qi, and H. Xu. 2018. "Collagen from marine biological sources and medical applications." *Chem Biodivers* 15 (5):e1700557. doi: 10.1002/cbdv.201700557.

Fu, Y., G. Li, X. Zhang, G. Xing, X. Hu, L. Yang, and D. Li. 2015. "Lipid extract from hardshelled mussel (Mytilus coruscus) improves clinical conditions of patients with rheumatoid arthritis: a randomized controlled trial." *Nutrients* 7 (1):625–45. doi: 10.3390/nu7010625.

Gomes, A. M., E. O. Kozlowski, V. H. Pomin, C. M. de Barros, J. L. Zaganeli, and M. S. Pavão. 2010. "Unique extracellular matrix heparan sulfate from the bivalve Nodipecten nodosus (Linnaeus, 1758) safely inhibits arterial thrombosis after photochemically induced endothelial lesion." *J Biol Chem* 285 (10):7312–23. doi: 10.1074/jbc.M109.091546.

Gómez-Guillén, M. C., B. Giménez, M. E. López-Caballero, and M. P. Montero. 2011. "Functional and bioactive properties of collagen and gelatin from alternative sources: A review." *Food Hydrocoll* 25 (8):1813–1827. doi: https://doi.org/10.1016/j.foodhyd.2011.02.007.

Gong, Fang, Mei-Fang Chen, Yuan-Yuan Zhang, Cheng-Yong Li, Chun-Xia Zhou, Peng-Zhi Hong, Sheng-Li Sun, and Zhong-Ji Qian. 2019. "A novel peptide from abalone (Haliotis discus hannai) to suppress metastasis and vasculogenic mimicry of tumor cells and enhance anti-tumor effect in vitro." *Mar Drugs* 17 (4):244.

Grienke, U., J. Silke, and D. Tasdemir. 2014. "Bioactive compounds from marine mussels and their effects on human health." *Food Chem* 142:48–60. doi: 10.1016/j.foodchem.2013.07.027.

Gulati, K., and K. M. Poluri. 2016. "Mechanistic and therapeutic overview of glycosaminoglycans: the unsung heroes of biomolecular signaling." *Glycoconj J* 33 (1):1–17. doi: 10.1007/s10719-015-9642-2.

Gurr, M. I., John L. Harwood, K. N. Frayn, Denis J. Murphy, R. H. Michell, and M. I. Gurr. 2016. *Lipids: Biochemistry, Biotechnology and Health.*

Halpern, G. M. 2000. "Anti-inflammatory effects of a stabilized lipid extract of Perna canaliculus (Lyprinol)." *Allerg Immunol (Paris)* 32 (7):272–278.

Han, J. H., J. S. Bang, Y. J. Choi, and S. Y. Choung. 2019. "Oral administration of oyster (Crassostrea gigas) hydrolysates protects against wrinkle formation by regulating the MAPK pathway in UVB-irradiated hairless mice." *Photochem Photobiol Sci* 18 (6):1436–1446. doi: 10.1039/c9pp00036d.

Han, M., Y. Zhao, W. Song, C. Wang, C. Mu, and R. Li. 2020. "Changes in microRNAs Expression Profile of Mimetic Aging Mice Treated with Melanin from Sepiella japonica Ink." *J Agric Food Chem* 68 (20):5616–5622. doi: 10.1021/acs.jafc.0c00291.

Harnedy, Pádraigín A., and Richard J. FitzGerald. 2012. "Bioactive peptides from marine processing waste and shellfish: a review." *J Funct Foods* 4 (1):6–24. doi: https://doi.org/10.1016/j.jff.2011.09.001.

Harris, W. S. 2004. "Fish oil supplementation: evidence for health benefits." *Cleve Clin J Med* 71 (3):208–10, 212, 215–8 passim. doi: 10.3949/ccjm.71.3.208.

He, Shudong, Yi Zhang, Hanju Sun, Ming Du, Jianlei Qiu, Mingming Tang, Xianbao Sun, and Beiwei Zhu. 2019. "Antioxidative peptides from proteolytic hydrolysates of false abalone (Volutharpa ampullacea perryi): characterization, identification, and molecular docking." *Mar Drugs* 17 (2):116.

Higashi, K., K. Takeda, A. Mukuno, Y. Okamoto, S. Masuko, R. J. Linhardt, and T. Toida. 2016. "Identification of keratan sulfate disaccharide at C-3 position of glucuronate of chondroitin sulfate from Mactra chinensis." *Biochem J* 473 (22):4145–4158. doi: 10.1042/bcj20160655.

Hovingh, Peter, and Alfred Linker. 1993. "Glycosaminoglycans in Anodonta californiensis, a Freshwater Mussel." *Biol Bull* 185 (2):263–276. doi: 10.2307/1542006.

Hu, X., L. Song, L. Huang, Q. Zheng, and R. Yu. 2012. "Antitumor effect of a polypeptide fraction from Arca subcrenata in vitro and in vivo." *Mar Drugs* 10 (12):2782–94. doi: 10.3390/md10122782.

Huang, F., Y. Jing, G. Ding, and Z. Yang. 2017. "Isolation and purification of novel peptides derived from Sepia ink: effects on apoptosis of prostate cancer cell PC-3." *Mol Med Rep* 16 (4):4222–4228. doi: 10.3892/mmr.2017.7068.

Huang, F., Z. Yang, D. Yu, J. Wang, R. Li, and G. Ding. 2012. "Sepia ink oligopeptide induces apoptosis in prostate cancer cell lines via caspase-3 activation and elevation of Bax/Bcl-2 ratio." *Mar Drugs* 10 (10):2153–65. doi: 10.3390/md10102153.

Hyung, J. H., C. B. Ahn, and J. Y. Je. 2018. "Blue mussel (Mytilus edulis) protein hydrolysate promotes mouse mesenchymal stem cell differentiation into osteoblasts through up-regulation of bone morphogenetic protein." *Food Chem* 242:156–161. doi: 10.1016/j.foodchem.2017.09.043.

Ifeanyi, D. Nwachukwu, and E. Aluko Rotimi. 2019. "Anticancer and antiproliferative properties of food-derived protein hydrolysates and peptides." *Journal of Food Bioactives* 7 (0). doi: 10.31665/JFB.2019.7194.

Ihn, H. J., J. A. Kim, S. Lim, S. H. Nam, S. H. Hwang, J. Lim, G. Y. Kim, 2019 Y. H. Choi, Y. J. Jeon, B. J. Lee, J. S. Bae, Y. H. Kim, and E. K. Park. 2019. "Fermented oyster extract prevents ovariectomy-induced bone loss and suppresses osteoclastogenesis." *Nutrients* 11 (6). doi: 10.3390/nu11061392.

Jia, W., Q. Peng, L. Su, X. Yu, C. W. Ma, M. Liang, X. Yin, Y. Zou, and Z. Huang. 2018. "Novel bioactive peptides from meretrix meretrix protect caenorhabditis elegans against free radical-induced oxidative stress through the stress response factor DAF-16/FOXO." *Mar Drugs* 16 (11). doi: 10.3390/md16110444.

Joshi, I., H. S. Mohideen, and R. A. Nazeer. 2021. "A Meretrix meretrix visceral mass derived peptide inhibits lipopolysaccharide-stimulated responses in RAW264.7 cells and adult zebrafish model." *Int Immunopharmacol* 90:107140. doi: 10.1016/j.intimp.2020.107140.

Joshi, I., and R. A. Nazeer. 2020a. "Anti-inflammatory potential of novel hexapeptide derived from Meretrix meretrix foot and its functional properties." *Amino Acids* 52 (10):1391–1401. doi: 10.1007/s00726-020-02899-0.

Joshi, I., and R. A. Nazeer. 2020b. "EGLLGDVF: a novel peptide from green mussel perna viridis foot exerts stability and anti-inflammatory effects on LPS-stimulated RAW264.7 cells." *Protein Pept Lett* 27 (9):851–859. doi: 10.2174/0929866527666200224111832.

Kang, Hee Kyoung, Hyung Ho Lee, Chang Ho Seo, and Yoonkyung Park. 2019. "Antimicrobial and immunomodulatory properties and applications of marine-derived proteins and peptides." *Mar Drugs* 17 (6):350. doi: 10.3390/md17060350.

Karthik, R., V. Manigandan, R. Saravanan, R. P. Rajesh, and B. Chandrika. 2016. "Structural characterization and in vitro biomedical activities of sulfated chitosan from Sepia pharaonis." *Int J Biol Macromol* 84:319–28. doi: 10.1016/j.ijbiomac.2015.12.030.

Khan, Bilal Muhammad, and Yang Liu. 2019. "Marine mollusks: food with benefits." *Compr Rev Food Sci Food Saf* 18 (2):548–564. doi: https://doi.org/10.1111/1541-4337.12429.

Kim, E. K., H. J. Joung, Y. S. Kim, J. W. Hwang, C. B. Ahn, Y. J. Jeon, S. H. Moon, and P. J. Park. 2012. "Purification of a novel anticancer peptide from enzymatic hydrolysate of Mytilus coruscus." *J Microbiol Biotechnol* 22 (10):1381–7. doi: 10.4014/jmb.1207.07015.

Kim, E. K., Y. S. Kim, J. W. Hwang, S. H. Kang, D. K. Choi, K. H. Lee, J. S. Lee, S. H. Moon, B. T. Jeon, and P. J. Park. 2013. "Purification of a novel nitric oxide inhibitory peptide derived from enzymatic hydrolysates of Mytilus coruscus." *Fish Shellfish Immunol* 34 (6):1416–20. doi: 10.1016/j.fsi.2013.02.023.

Kimura, S., Y. Takema, and M. Kubota. 1981. "Octopus skin collagen. Isolation and characterization of collagen comprising two distinct alpha chains." *J Biol Chem* 256 (24):13230–4.

Le, Cheng Foh, Chee-Mun Fang, and Shamala Sekaran. 2017. "Intracellular targeting mechanisms by antimicrobial peptides." *Antimicrobi Agents Chemother* 61:AAC.02340-16. doi: 10.1128/AAC.02340-16.

Li, F., P. Luo, and H. Liu. 2018. "A potential adjuvant agent of chemotherapy: sepia ink polysaccharides." *Mar Drugs* 16 (4). doi: 10.3390/md16040106.

Li, H., M. Giovanna Parisi, Nicolò Parrinello, Matteo Cammarata, and Philippe Roch. 2011. "Molluscan antimicrobial peptides, a review from activity-based evidences to computer-assisted sequences."

Li, J., C. Gong, Z. Wang, R. Gao, J. Ren, X. Zhou, H. Wang, H. Xu, F. Xiao, Y. Cao, and Y. Zhao. 2019. "Oyster-derived zinc-binding peptide modified by plastein reaction via zinc chelation promotes the intestinal absorption of zinc." *Mar Drugs* 17 (6). doi: 10.3390/md17060341.

Li, Li, Heng Li, Jianying Qian, Yongfeng He, Jialin Zheng, Zhenming Lu, Zhenghong Xu, and Jinsong Shi. 2015. "Structural and immunological activity characterization of a polysaccharide isolated from meretrix meretrix linnaeus." *Mar Drugs* 14 (1):6–6. doi: 10.3390/md14010006.

Li, W., S. Ye, Z. Zhang, J. Tang, H. Jin, F. Huang, Z. Yang, Y. Tang, Y. Chen, G. Ding, and F. Yu. 2019. "Purification and characterization of a novel pentadecapeptide from protein hydrolysates of cyclina sinensis and its immunomodulatory effects on RAW264.7 cells." *Mar Drugs* 17 (1). doi: 10.3390/md17010030.

Liu, P., X. Lan, M. Yaseen, S. Wu, X. Feng, L. Zhou, J. Sun, A. Liao, D. Liao, and L. Sun. 2019. "Purification, characterization and evaluation of inhibitory mechanism of ACE inhibitory peptides from pearl oyster (Pinctada fucata martensii) meat protein hydrolysate." *Mar Drugs* 17 (8):463. doi: 10.3390/md17080463. PMID: 31398788; PMCID: PMC6723713.

Lozano, Azalia, Shela Gorinstein, Eduardo Espitia-Rangel, Gloria Dávila-Ortiz, and A. Martinez-Ayala. 2018. "Plant sources, extraction methods, and uses of squalene." *Int J Agron* 2018:1–13. doi: 10.1155/2018/1829160.

Melnick, S. C. 1958. "Occurrence of collagen in the phylum mollusca." *Nature* 181 (4621):1483–1483. doi: 10.1038/1811483a0.

Merdzhanova, Albena, Diana Dobreva, Mona Stancheva, and Lubomir Makedonski. 2014. "Fat soluble vitamins and fatty acid composition of wild Black sea mussel, rapana and shrimp." *Ovidius Univ Ann Chem* 25:pp. 15–23. doi: 10.2478/auoc-2014-0003.

Micera, M., A. Botto, F. Geddo, S. Antoniotti, C. M. Bertea, R. Levi, M. P. Gallo, and G. Querio. 2020. "Squalene: more than a step toward Sterols." *Antioxidants (Basel)* 9 (8). doi: 10.3390/antiox9080688.

Miller, M. R., L. Pearce, and B. I. Bettjeman. 2014. "Detailed distribution of lipids in Greenshell™ mussel (Perna canaliculus)." *Nutrients* 6 (4):1454–74. doi: 10.3390/nu6041454.

Miralles, B., L. Amigo, and I. Recio. 2018. "Critical review and perspectives on food-derived antihypertensive peptides." *Journal of Agricultural and Food Chemistry* 66 (36):9384–9390.

Mizuta, Shoshi, Tomoyuki Miyagi, Tohru Nishimiya, and Reiji Yoshinaka. 2004. "Partial characterization of collagen in several bivalve molluscs." *Food Chem* 87 (1):83–88. doi: https://doi.org/10.1016/j.foodchem.2003.10.021.

Molagoda, I. M. N., Wahm Karunarathne, Y. H. Choi, E. K. Park, Y. J. Jeon, B. J. Lee, C. H. Kang, and G. Y. Kim. 2019. "Fermented oyster extract promotes osteoblast differentiation by activating the Wnt/β-catenin signaling pathway, leading to bone formation." *Biomolecules* 9 (11). doi: 10.3390/biom9110711.

Murphy, K. J., N. J. Mann, and A. J. Sinclair. 2003. "Fatty acid and sterol composition of frozen and freeze-dried New Zealand Green Lipped Mussel (Perna canaliculus) from three sites in New Zealand." *Asia Pac J Clin Nutr* 12 (1):50–60.

Murphy, Karen J., Ben D. Mooney, Neil J. Mann, Peter D. Nichols, and Andrew J. Sinclair. 2002. "Lipid, FA, and sterol composition of New Zealand green lipped mussel (Perna canaliculus) and Tasmanian blue mussel (Mytilus edulis)." *Lipids* 37 (6):587–595. doi: 10.1007/s11745-002-0937-8.

Muthuvel, Arumugam, Hari Garg, Thangappan Ajithkumar, and Annaian Shanmugam. 2009. "Antiproliferative heparin (glycosaminoglycans) isolated from giant clam (Tridacna maxima) and green mussel (Perna viridis)." 8:2394–2396.

Narayan, B. H., N. Tatewaki, V. V. Giridharan, H. Nishida, and T. Konishi. 2010. "Modulation of doxorubicin-induced genotoxicity by squalene in Balb/c mice." *Food Funct* 1 (2):174–9. doi: 10.1039/c0fo00102c.

Narayan Bhilwade, H., N. Tatewaki, T. Konishi, M. Nishida, T. Eitsuka, H. Yasui, O. Inanami, O. Handa, Y. Naito, N. Ikekawa, and H. Nishida. 2019. "The adjuvant effect of squalene, an active ingredient of functional foods, on doxorubicin-treated allograft mice." *Nutr Cancer* 71 (7):1153–1164. doi: 10.1080/01635581.2019.1597900.

Natarajan, Sithranga Boopathy, Yon-Suk Kim, Jin-Woo Hwang, and Pyo-Jam Park. 2016. "Immunomodulatory properties of shellfish derivatives associated with human health." *RSC Advances* 6 (31):26163–26177. doi: 10.1039/C5RA26375A.

Nelson, D. L., and M. M. Cox. 2012. *Lehninger Principles of Biochemistry*. 6th Edition, Macmillan Learning.

Odeleye, T., W. L. White, and J. Lu. 2019. "Extraction techniques and potential health benefits of bioactive compounds from marine molluscs: a review." *Food Funct* 10 (5):2278–2289. doi: 10.1039/c9fo00172g.

Oh, Y., C. B. Ahn, K. H. Nam, Y. K. Kim, N. Y. Yoon, and J. Y. Je. 2019. "Amino acid composition, antioxidant, and cytoprotective effect of blue mussel (Mytilus edulis) hydrolysate through the inhibition of caspase-3 activation in oxidative stress-mediated endothelial cell injury." *Mar Drugs* 17 (2). doi: 10.3390/md17020135.

Oh, Yunok, Chang-Bum Ahn, Jun-Ho Hyung, and Jae-Young Je. 2019. "Two novel peptides from ark shell protein stimulate osteoblast differentiation and rescue ovariectomy-induced bone loss." *Toxicol Appl Pharmacol* 385:114779. doi: https://doi.org/10.1016/j.taap.2019.114779.

Phillips, K. M., D. M. Ruggio, J. Exler, and K. Y. Patterson. 2012. "Sterol composition of shellfish species commonly consumed in the United States." *Food Nutr Res* 56. doi: 10.3402/fnr.v56i0.18931.

Pikkarainen, J., J. Rantanen, M. Vastamäki, K. Lampliaho, A. Kari, and E. Kulonen. 1968. "On collagens of invertebrates with special reference to mytilus edulis." *Euro J Biochem* 4 (4):555–560. doi: https://doi.org/10.1111/j.1432-1033.1968.tb00248.x.

Pomin, V. H. 2014. "Marine medicinal glycomics." *Front Cell Infect Microbiol* 4:5. doi: 10.3389/fcimb.2014.00005.

Qian, Z. J., W. K. Jung, H. G. Byun, and S. K. Kim. 2008. "Protective effect of an antioxidative peptide purified from gastrointestinal digests of oyster, Crassostrea gigas against free radical induced DNA damage." *Bioresour Technol* 99 (9):3365–71. doi: 10.1016/j.biortech.2007.08.018.

Qiao, M., M. Tu, Z. Wang, F. Mao, H. Chen, L. Qin, and M. Du. 2018. "Identification and antithrombotic activity of peptides from blue mussel (mytilus edulis) protein." *Int J Mol Sci* 19 (1). doi: 10.3390/ijms19010138.

Rahman, M. Azizur. 2019. "Collagen of extracellular matrix from marine invertebrates and its medical applications." *Marine Drugs* 17 (2):118.

Ruocco, N., S. Costantini, S. Guariniello, and M. Costantini. 2016. "Polysaccharides from the marine environment with pharmacological, cosmeceutical and nutraceutical potential." *Molecules* 21 (5). doi: 10.3390/molecules21050551.

Sahena, F., I. S. M. Zaidul, S. Jinap, N. Saari, H. A. Jahurul, K. A. Abbas, and N. A. Norulaini. 2009. "PUFAs in fish: extraction, fractionation, importance in health." *Compr Rev Food Sci Food Safe* 8 (2):59–74. doi: https://doi.org/10.1111/j.1541-4337. 2009.00069.x.

Salvatore, L., N. Gallo, M. L. Natali, L. Campa, P. Lunetti, M. Madaghiele, F. S. Blasi, A. Corallo, L. Capobianco, and A. Sannino. 2020. "Marine collagen and its derivatives: Versatile and sustainable bio-resources for healthcare." *Mater Sci Eng C Mater Biol Appl* 113:110963. doi: 10.1016/j.msec.2020.110963.

Saravanan, R., S. Vairamani, and A. Shanmugam. 2010. "Glycosaminoglycans from marine clam Meretrix meretrix (Linne.) are an anticoagulant." *Prep Biochem Biotechnol* 40 (4):305–315. doi: 10.1080/10826068.2010.488998.

Sasisekharan, R., R. Raman, and V. Prabhakar. 2006. "Glycomics approach to structure-function relationships of glycosaminoglycans." *Annu Rev Biomed Eng* 8:181–231. doi: 10.1146/annurev.bioeng.8.061505.095745.

Sathyan, Naveen, Rosamma Philip, E. R. Chaithanya, and P. R. Anil Kumar. 2012. "Identification and molecular characterization of molluskin, a histone-H2A-derived antimicrobial peptide from molluscs." *ISRN molecular biology* 2012:219656–219656. doi: 10.5402/2012/219656.

Sato, K. 2017. "The presence of food-derived collagen peptides in human body-structure and biological activity." *Food Funct* 8 (12):4325–4330. doi: 10.1039/c7fo01275f.

Seedevi, P., M. Moovendhan, S. Vairamani, and A. Shanmugam. 2017. "Mucopolysaccharide from cuttlefish: purification, chemical characterization and bioactive potential." *Carbohydr Polym* 167:129–135. doi: 10.1016/j.carbpol.2017.03.028.

Senevirathne, M., and S. K. Kim. 2012. "Development of bioactive peptides from fish proteins and their health promoting ability." *Adv Food Nutr Res* 65:235–48. doi: 10.1016/b978-0-12-416003-3.00015-9.

Shanmugam, Annaian, Thangaraj Amalraj, Chendur Palpandi, and Balasubramanian Thangavel. 2008. "Antimicrobial activity of sulfated mucopolysaccharides [heparin and heparin-like glycosaminoglycans (GAGs)] from cuttlefish euprymna berryi sasaki, 1929." *Trends Tech Sci Res* 3:97–102. doi: 10.3923/tasr.2008.97.102.

Siregar, A. S., M. M. Nyiramana, E. J. Kim, E. J. Shin, M. S. Woo, J. M. Kim, J. H. Kim, D. K. Lee, J. R. Hahm, H. J. Kim, C. W. Kim, N. G. Kim, S. H. Park, Y. J. Choi, S. S. Kang, S. G. Hong, J. Han, and D. Kang. 2020. "Dipeptide YA is responsible for the positive effect of oyster hydrolysates on alcohol metabolism in single ethanol binge rodent models." *Mar Drugs* 18 (10). doi: 10.3390/md18100512.

Sivakumar, Pitchumani, Lonchin Suguna, and Gowri Chandrakasan. 2003. "Similarity between the major collagens of cuttlefish cranial cartilage and cornea." *Comp Biochem Physiol B, Biochem Mol Biol Comp Biochem Phys B* 134 (1):171–180. doi: 10.1016/s1096-4959(02)00224-5.

Smith, Theresa J. 2000. "Squalene: potential chemopreventive agent." *Expert Opin Investig Drugs* 9 (8):1841–1848. doi: 10.1517/13543784.9.8.1841.

Soliman, A. M., S. R. Fahmy, and S. A. El-Abied. 2015. "Anti-neoplastic activities of Sepia officinalis ink and Coelatura aegyptiaca extracts against Ehrlich ascites carcinoma in Swiss albino mice." *Int J Clin Exp Pathol* 8 (4):3543–55.

Stancheva, Mona, Albena Merdzhanova, and Diana A. Dobreva. 2017. "Fat soluble vitamins, cholesterol, and fatty acid composition of wild and farmed black mussel (mytilus galloprovincialis) consumed in Bulgaria." *J Aquat Food Prod Technol* 26 (2):181–191. doi: 10.1080/10498850.2015.1108378.

Suarez-Jimenez, G. M., A. Burgos-Hernandez, and J. M. Ezquerra-Brauer. 2012. "Bioactive peptides and depsipeptides with anticancer potential: sources from marine animals." *Mar Drugs* 10 (5):963–86. doi: 10.3390/md10050963.

Suárez-Jiménez, G. Miroslava, Armando Burgos-Hernández, Wilfrido Torres-Arreola, Carmen M. López-Saiz, Carlos A. Velázquez Contreras, and J. Marina Ezquerra-Brauer. 2019. "Bioactive peptides from collagen hydrolysates from squid (Dosidicus gigas) by-products fractionated by ultrafiltration." *Int J Food Sci Tech* 54 (4):1054–1061. doi: https://doi.org/10.1111/ijfs.13984.

Sudhakar, Sekar, and Rasool Abdul Nazeer. 2015. "Structural characterization of an Indian squid antioxidant peptide and its protective effect against cellular reactive oxygen species." *J Funct Foods* 14:502–512. doi: https://doi.org/10.1016/j.jff.2015.02.028.

Tabakaeva, Oksana Vatslavovna, Anton Vadimovich Tabakaev, and Wojciech Piekoszewski. 2018. "Nutritional composition and total collagen content of two commercially important edible bivalve molluscs from the Sea of Japan coast." *J Food Sci Tech* 55 (12):4877–4886. doi: 10.1007/s13197-018-3422-5.

Takema, Yoshinori, and Shigeru Kimura. 1982. "Two genetically distinct molecular species of octopus muscle collagen." *Biochimica et Biophysica Acta (BBA)—Protein Struct Mol Enzymol* 706 (1):123–128. doi: https://doi.org/10.1016/0167-4838(82)90382-X.

Tan, K., H. Ma, S. Li, and H. Zheng. 2020. "Bivalves as future source of sustainable natural omega-3 polyunsaturated fatty acids." *Food Chem* 311:125907. doi: 10.1016/j.foodchem.2019.125907.

Tang, Y., Y. Cui, A. De Agostini, and L. Zhang. 2019. "Biological mechanisms of glycan- and glycosaminoglycan-based nutraceuticals." *Prog Mol Biol Transl Sci* 163:445–469. doi: 10.1016/bs.pmbts.2019.02.012.

Teshima, S., and A. Kanazawa. 1974. "Biosynthesis of sterols in abalone, Haliotis gurneri, and mussel, Mytilus edulis." *Comp Biochem Physiol B* 47 (3):555–61. doi: 10.1016/0305-0491(74)90004-2.

Urich, Klaus. 1994. "Sterols and steroids." In *Comparative Animal Biochemistry*, edited by Klaus Urich, 624–656. Berlin, Heidelberg: Springer Berlin Heidelberg.

Valcarcel, Jesus, Ramon Novoa-Carballal, Ricardo I. Pérez-Martín, Rui L. Reis, and José Antonio Vázquez. 2017. "Glycosaminoglycans from marine sources as therapeutic agents." *Biotech Adv* 35 (6):711–725. doi: 10.1016/j.biotechadv.2017.07.008.

Valtchev, Peter. 2018. "Study on novel antibacterial and antiviral compounds from abalone as an important marine mollusc." *J Aquacu Mar Bio* 7. doi: 10.15406/jamb.2018.07.00200.

Venkatesan, J., S. Anil, S. K. Kim, and M. S. Shim. 2017. "Marine fish proteins and peptides for cosmeceuticals: a review." *Mar Drugs* 15 (5). doi: 10.3390/md15050143.

Vijayabaskar, P., T. Balasubramanian, and S. T. Somasundaram. 2008. "Low-molecular weight molluscan glycosaminoglycan from bivalve Katelysia opima (Gmelin)." *Methods Find Exp Clin Pharmacol* 30 (3):175–80. doi: 10.1358/mf.2008.30.3.1159654.

Vijayabaskar, P., and S. T. Somasundaram. 2012. "Studies on molluscan glycosaminoglycans (GAG) from backwater clam Donax cuneatus (Linnaeus)." *Asian Pac J Trop Biomed* 2 (2, Supplement):S519-S525. doi: https://doi.org/10.1016/S2221-1691(12)60265-2.

Volpi, Nicola, and Francesca Maccari. 2005. "Glycosaminoglycan composition of the large freshwater mollusc bivalve anodonta." *Biomacromolecules* 6 (6):3174–3180. doi: 10.1021/bm0505033.

Volpi, Nicola, and Francesca Maccari. 2008. "Structural characterization and antithrombin activity of dermatan sulfate purified from marine clam Scapharca inaequivalsis." *Glycobiology* 19 (4):356–367. doi: 10.1093/glycob/cwn140.

Voogt, P. A. 1973. "Investigations of the capacity of synthesizing 3 beta-sterols in mollusca. X. Biosynthesis and composition of 3 beta-sterols in Cephalopoda." *Arch Int Physiol Biochim* 81 (3):401–7. doi: 10.3109/13813457309073391.

Wakimoto, Toshiyuki, Hikaru Kondo, Hirohiko Nii, Kaori Kimura, Yoko Egami, Yusuke Oka, Masae Yoshida, Eri Kida, Yiping Ye, Saeko Akahoshi, Tomohiro Asakawa, Koichi Matsumura, Hitoshi Ishida, Haruo Nukaya, Kuniro Tsuji, Toshiyuki Kan, and Ikuro Abe. 2011. "Furan fatty acid as an anti-inflammatory component from the green-lipped mussel Perna canaliculus." *Pro Natl Acad Sci* 108 (42):17533–17537. doi: 10.1073/pnas.1110577108.

Wallace, R. L., and W. K. Taylor. 2003. *Invertebrate Zoology: A Laboratory Manual.* 6th Edition, Prentice Hall.

Wang, B., L. Li, C. F. Chi, J. H. Ma, H. Y. Luo, and Y. F. Xu. 2013. "Purification and characterisation of a novel antioxidant peptide derived from blue mussel (Mytilus edulis) protein hydrolysate." *Food Chem* 138 (2–3):1713–9. doi: 10.1016/j.foodchem.2012.12.002.

Wang, L. C., L. Q. Di, J. S. Li, L. H. Hu, J. M. Cheng, and H. Wu. 2019. "Elaboration in type, primary structure, and bioactivity of polysaccharides derived from mollusks." *Crit Rev Food Sci Nutr* 59 (7):1091–1114. doi: 10.1080/10408398.2017.1392289.

Wang, L., L. Chen, J. Li, L. Di, and H. Wu. 2018. "Structural elucidation and immune-enhancing activity of peculiar polysaccharides fractioned from marine clam Meretrix meretrix (Linnaeus)." *Carbohydr Polym* 201:500–513. doi: 10.1016/j.carbpol.2018.08.106.

Wang, L., Y. Yang, H. Y. Tan, S. Li, and Y. Feng. 2020. "Protective actions of acidic hydrolysates of polysaccharide extracted from mactra veneriformis against chemical-induced acute liver damage." *Front Pharmacol* 11:446. doi: 10.3389/fphar.2020.00446.

Wang, X., H. Yu, R. Xing, and P. Li. 2017. "Characterization, preparation, and purification of marine bioactive peptides." *Biomed Res Int* 2017:9746720. doi: 10.1155/2017/9746720.

Wang, Y. K., H. L. He, G. F. Wang, H. Wu, B. C. Zhou, X. L. Chen, and Y. Z. Zhang. 2010. "Oyster (Crassostrea gigas) hydrolysates produced on a plant scale have antitumor activity and immunostimulating effects in BALB/c mice." *Mar Drugs* 8 (2):255–68. doi: 10.3390/md8020255.

Wu, J., X. Cai, M. Tang, and S. Wang. 2019. "Novel calcium-chelating peptides from octopus scraps and their corresponding calcium bioavailability." *J Sci Food Agric* 99 (2):536–545. doi: 10.1002/jsfa.9212.

Xu, Z., H. Chen, F. Fan, P. Shi, M. Tu, S. Cheng, Z. Wang, and M. Du. 2019. "Bone formation activity of an osteogenic dodecapeptide from blue mussels (Mytilus edulis)." *Food Funct* 10 (9):5616–5625. doi: 10.1039/c9fo01201j.

Xu, Z., F. Zhao, H. Chen, S. Xu, F. Fan, P. Shi, M. Tu, Z. Wang, and M. Du. 2019. "Nutritional properties and osteogenic activity of enzymatic hydrolysates of proteins from the blue mussel (Mytilus edulis)." *Food Funct* 10 (12):7745–7754. doi: 10.1039/c9fo01656b.

Yang, X. R., Y. T. Qiu, Y. Q. Zhao, C. F. Chi, and B. Wang. 2019. "Purification and characterization of antioxidant peptides derived from protein hydrolysate of the marine bivalve mollusk tergillarca granosa." *Mar Drugs* 17 (5). doi: 10.3390/md17050251.

Yao, H. T., P. F. Lee, C. K. Lii, Y. T. Liu, and S. H. Chen. 2018. "Freshwater clam extract reduces liver injury by lowering cholesterol accumulation, improving dysregulated cholesterol synthesis and alleviating inflammation in high-fat, high-cholesterol and cholic acid diet-induced steatohepatitis in mice." *Food Funct* 9 (9):4876–4887. doi: 10.1039/c8fo00851e.

Yu, F., Z. Zhang, L. Luo, J. Zhu, F. Huang, Z. Yang, Y. Tang, and G. Ding. 2018. "Identification and molecular docking study of a novel angiotensin-I converting enzyme inhibitory peptide derived from enzymatic hydrolysates of cyclina sinensis." *Mar Drugs* 16 (11). doi: 10.3390/md16110411.

Yu, Fangmiao, Yaru Zhang, Lei Ye, Yunping Tang, Guofang Ding, Xiaojun Zhang, and Zuisu Yang. 2018. "A novel anti-proliferative pentapeptide (ILYMP) isolated from Cyclina sinensis protein hydrolysate induces apoptosis of DU-145 prostate cancer cells." *Mol Med Rep* 18 (1):771–778. doi: 10.3892/mmr.2018.9019.

Yu, X., Q. Su, T. Shen, Q. Chen, Y. Wang, and W. Jia. 2020. "Antioxidant peptides from sepia esculenta hydrolyzate attenuate oxidative stress and fat accumulation in caenorhabditis elegans." *Mar Drugs* 18 (10). doi: 10.3390/md18100490.

Zasloff, Michael. 2002. "Antimicrobial peptides of multicellular organisms." *Nature* 415 (6870):389–395. doi: 10.1038/415389a.

Zhang, S. S., L. W. Han, Y. P. Shi, X. B. Li, X. M. Zhang, H. R. Hou, H. W. Lin, and K. C. Liu. 2018. "Two novel multi-functional peptides from meat and visceral mass of marine snail neptunea arthritica cumingii and their activities in vitro and in vivo." *Mar Drugs* 16 (12). doi: 10.3390/md16120473.

Zhang, Z., G. Su, F. Zhou, L. Lin, X. Liu, and M. Zhao. 2019. "Alcalase-hydrolyzed oyster (Crassostrea rivularis) meat enhances antioxidant and aphrodisiac activities in normal male mice." *Food Res Int* 120:178–187. doi: 10.1016/j.foodres.2019.02.033.

Zhang, Z., L. Sun, G. Zhou, P. Xie, and J. Ye. 2017. "Sepia ink oligopeptide induces apoptosis and growth inhibition in human lung cancer cells." *Oncotarget* 8 (14):23202–23212. doi: 10.18632/oncotarget.15539.

Zhang, Z., F. Zhou, X. Liu, and M. Zhao. 2018. "Particulate nanocomposite from oyster (Crassostrea rivularis) hydrolysates via zinc chelation improves zinc solubility and peptide activity." *Food Chem* 258:269–277. doi: 10.1016/j.foodchem.2018.03.030.

Zhuang, J., C. J. Coates, H. Zhu, P. Zhu, Z. Wu, and L. Xie. 2015. "Identification of candidate antimicrobial peptides derived from abalone hemocyanin." *Dev Comp Immunol* 49 (1):96–102. doi: 10.1016/j.dci.2014.11.008.

Index

Note: numbers in **bold** indicate a table. Numbers in *italics* indicate a figure.

K, 111, 121, 123, **125**
K1, 123
mollusks, 110–111
seaweed-derived, 49–53

W

wild capture, 2
    adult sea urchins, 82
    broodstock of sea cucumber, 73
    juvenile sea urchins, 77

X

Xanthomanes vesicatoria, 32
Xanthophyllomyces, 37
xanthophylls, 43
Xylaria spp., 36
Xylated glycosaminoglycan, **128**
xylose, 31
    d-xylose, 48

Y

yellow color, *see* carotenoid pigments
Yellowfin sole (*Limanda aspera*), 5

Z

(Z)-5-(Hydroxymethyl)-2-(6')-
    methylhept-2-en-2'-yl)-phenol,
    33, 36
zanamivir, 36
Zonaria angustata, 84
Zostera noltii, 74
zooplankton, 95–105
    ASX and, 100
    digestion of, 103
    fatty acids and amino acids of, 98–105
    mesozooplankton, 99
    microzooplankton, 99
    see also Calanus oil
Zygomycota, 31

For Product Safety Concerns and Information please contact our EU
representative GPSR@taylorandfrancis.com
Taylor & Francis Verlag GmbH, Kaufingerstraße 24, 80331 München, Germany